T0321199

THE LOGICAL FOUNDATIONS
OF THE
MARXIAN THEORY OF VALUE

SYNTHESE LIBRARY

STUDIES IN EPISTEMOLOGY,

LOGIC, METHODOLOGY, AND PHILOSOPHY OF SCIENCE

VOLUME 223

THE LOGICAL FOUNDATIONS
OF THE
MARXIAN THEORY OF VALUE

ADOLFO GARCÍA DE LA SIENRA

Institute of Philosophical Research,
Universidad Nacional Autónoma de México

KLUWER ACADEMIC PUBLISHERS

DORDRECHT / BOSTON / LONDON

Library of Congress Cataloging-in-Publication Data

```
García de la Sienra, Adolfo
    The logical foundations of the Marxian theory of value / Adolfo
García de la Sienra.
        p.    cm. -- (Synthese library ; v. 223)
    Includes bibliographical references and index.
    ISBN 0-7923-1778-5 (alk. paper)
    1. Surplus value. 2. Labor theory of value. 3. Marxian
economics.    I. Title. II. Series.
HB206.G34   1992
335.4'12--dc20                                        92-14028
```

ISBN 0-7923-1778-5

Published by Kluwer Academic Publishers,
P.O. Box 17, 3300 AA Dordrecht, The Netherlands.

Kluwer Academic Publishers incorporates
the publishing programmes of
D. Reidel, Martinus Nijhoff, Dr W. Junk and MTP Press.

Sold and distributed in the U.S.A. and Canada
by Kluwer Academic Publishers,
101 Philip Drive, Norwell, MA 02061, U.S.A.

In all other countries, sold and distributed
by Kluwer Academic Publishers Group,
P.O. Box 322, 3300 AH Dordrecht, The Netherlands.

Printed on acid-free paper

Printed in the Netherlands

To Carolina

TABLE OF CONTENTS

INTRODUCTION

After the impressive collapse of the Soviet Union and the Eastern European socialist countries, it is more pertinent than ever to recover the scientific legacy of Karl Marx. This legacy is mainly (if not exclusively) constituted by his work in the field of economic theory. Marx's economic theory was intended by his author as a scientific objective theory about the nature of capitalist economies, a theory that was going to serve as the foundation of the critique of bourgeois political economy. His "laws" about the demise of capitalism, like the tendency of the profit rate to fall or the law of the cyclical crises, have been shown to hold under certain conditions but not in general. At any rate, it is likely that had not the industrialized countries changed the situation of the working class, and allowed some intervention of the State in the economy (especially after the Great Depression), capitalism would have hardly survived, even though it is impossible to guess what kind of regime would have been instaurated in its place.

The present book is concerned with the very foundations of Marx's economics, hence with the very foundations of his scientific legacy. I hope that after reading the book the reader will be convinced that Marx's scientific work was indeed serious and that this is the time to recover it as an important paradigm in

1

scientific research. I think that the reader will be convinced that Marx's economic theory is no less serious and mathematically tractable than, say, general equilibrium theory. Quite another question is whether Marx's economic theory constitutes by itself a critique of capitalism. Some authors would say yes since, for one thing, the Fundamental Marxian Theorem is one of its central results. This theorem asserts that the exploitation of the workers is both necessary and sufficient for capitalist profit. These authors would claim that —in consequence— capitalism is stripping the worker part of his labor, but this view —which I am not despising at all— can be adopted in a very naive way, without taking into account many other factors like the risk taken by entrepeneurs, the cost of creating jobs, the need to have an incentive for investment, the creativity required of entrepeneurs, and so on. In the present book I do not consider these questions, but only apply myself to put a solid logical foundation for further development of the labor theory of value.

The labor theory of value as presented by Marx in *Capital* began to receive mathematical treatment toward the end of the nineteenth century, and was reformulated with some completion toward the beginning of the fifties, thanks mainly to the work of Leontief. The mathematical reconstruction —or formulation— of Marx's original theory I shall call 'the prototype'. In the first part of the book I present a rather detailed formulation of the prototype, proving the most basic theorems. The chapter begins with a detailed discussion of the concept of value in Marx's *Capital*. It is shown thereby that there is an ambiguity in Marx's conception of value. In some passages, Marx insists that the magnitude of value must be determined independently of market relations, solely in the sphere of production, whereas in others Marx writes as if the market were the factor reducing heterogeneous to abstract labor. In *Capital*, as well as in the usual Marxist literature, the first interpretation has prevailed over the other and, accordingly, the prototype is built upon this interpretation.

One of the most outstanding exponents of the prototype has been Morishima (1973). My own formulation of the prototype owes a lot to the one provided by Morishima, but there are some differences that have mainly to do with the methods of proof. The idealized economy described by the prototype I call a 'simple Marxian economy'. The main result of the chapter is the proof that the Law of Value, namely that prices are proportional to values, holds only if all industries in the economy have the same value composition of capital.

In the second chapter I proceed to discuss in some detail the foundational problems of the theory of value. These problems arise from the difficulties faced by Marx to clarify in what sense "value regulates prices". After reviewing all the attempts made by Marx to solve this problem (throughout volume 3 of *Capital*), I discuss the criticisms advanced by Böhm-Bawerk against such attempts, in order to conclude that these criticisms are sound. Indeed, this book can be seen as an attempt to provide new foundations for the theory of value, foundations that are no longer subject to Böhm-Bawerk devastating criticisms. This attempt is deeply related to the task of generalizing the prototype. In the same chapter, I present eight restrictive assumptions built into the prototype that are candidates to be eliminated. The chapter ends presenting a concise history of the efforts made to get rid of such assumptions. It is shown that the main problem is to allow heterogeneous labor in a rather general setting of production sets.

Chapter 3 is a revision of formal techniques. I present the outlines of a system of mathematical logic, up to the concept of a model, in order to discuss the need of introducing another concept of model more akin to Bourbaki's concept of structure. In the second part I discuss the problem of the relation between abstract mathematical models and scientific objects. As a way of approaching this problem, I present the concept of fundamental measurement together with an example that illustrates clearly —or so I hope— the idea of a fundamental measurement. The measurement I provide afterward —the measurement of abstract labor— is not fundamental, because the structure to be measured already is a mathematical one, of course having a very specific economic meaning. But the techniques explained and exemplified in the chapter hold also for this case, except that the structure to be measured is not "ontological".

Chapter four is a rather long chapter in which I discuss Hegel's dialectical method and its relation to Marx's dialectical logic. After presenting in some detail Hegel's method, I discuss Marx's "inversion" of Hegel in order to show that his misapplication of Hegel's method might be responsible for his severing the determination of value from the market relations. In order to proceed to a new formulation of dialectic, I analyze the theological framework of Hegel's dialectic in order to conclude that Hegel's concept of Spirit (*Geist*) is not acceptable on theological and metaphysical grounds, since it implies a particularly strong form of Averroism. After discussing Hegel's theory of universals, I discuss a little the root of the so-called "problem of incommensurability", which indeed is nothing but a pseudoproblem from this point of view. In the last section I develop in some detail this modified Hegelian view as a theory of science with special reference to the axiomatic method.

In chapter 5 I claim that Marx's decision to sever value from exchange is a dialectical mistake. This is established as a corollary of the main result of chapter 1, because the proposition that the transformation problem is solvable is logically inconsistent with the proposition that the organic composition of capital is *not* the same for all industries. After reviewing other passages of Marx's work, especially in the *A Contribution to a Critique of Political Economy*, where Marx suggests that value is a result of the exchange process and not a ready made prerequisite, I adopt this different view. This view had been held only by Isaak Illich Rubin in the twenties and, more recently and using mathematical techniques, by Professor Ulrich Krause (University of Bremen). After considering the important work of Krause's, I introduce a formal definition of abstract labor and then proceed to provide a more general proof of the existence of a representation for abstract labor. This representation or measurement of abstract labor is what we call 'value'. The Law of Value results as a corollary of the representation theorem for abstract labor.

The general axioms of the labor theory of value are introduced in chapter six. This theory is thereby axiomatized through the definition of the set-theoretical predicate 'Marxian capitalist economy'. The fundamental law of the theory is not the Law of Value, but one asserting that the behavior of the capitalist firms consists of maximizing their profit. Nevertheless, the Law of Value is a consequence of this axiom. Another consequence of the same is the Fundamental Marxian Theorem, according to which the exploitation of the workers is both necessary and sufficient for capitalist profit.

Chapter 7 is by far the most technical of the book. In this chapter I define the concept of a reproducible global decision, i.e. a global production process chosen by the firms which can reproduce itself in the sense of being able to reproduce all the wage and capital goods its operation requires. The existence of a reproducible global decision is actually equivalent to the existence of a special kind of competitive equilibrium —that I label a 'Marxian competitive equilibrium'. Almost all of the chapter is devoted to prove the existence of this equilibrium, as a means of proving the existence of the reproducible global process.

In the last chapter of the book I return to the prototype to see how the foundational problems of the theory have been solved. In fact, the Law of Value holds in Leontief economies in the strongest possible form: if the equilibrium price system in the sense of chapter 7 is profitable for all the firms, then this price system is actually positive and equal to the the unique price system at which the profit rate is positive; moreover, the labor-values are equal to this price system. In other words, in chapter 8 the Law of Value means that labor-values are the (unique) equilibrium prices. In this form, the "dialectical contradiction" caused by Marx's severing of value from exchange is solved and the theory of value is seen to have a solid scientific foundation.

Chapter 1

THE PROTOTYPE OF MARX'S
LABOR THEORY OF VALUE

In the first two volumes of *Capital*,[1] Karl Marx developed in a rather informal fashion a theory of value. It is known that this theory of value owes a lot to Ricardo's labor theory of value, but I shall not be concerned in the present book with establishing in what respects Marx's theory is similar or different from that of Ricardo's. Instead, I shall take Marx's theory as point of departure of the history of a theory of value that everybody shall agree in calling 'the Marxian theory of value' (MTV, for short).

One of the theoretical aims pursued by Marx in *C* was to prove that the source of the capitalist's profits is nothing but that unpaid part of the worker's labor which he calls surplus-value. In order to prove this, it is obviously necessary to define first a quantitative concept of value, since surplus-value is just the difference in value between the goods the worker can buy with his salary and those he produced to earn that salary. Hence, any serious attempt to make sense of Marx's MTV as formulated in the first two volumes of *C must* show in the first place that value, namely as a quantitative concept, is indeed well defined. This is also required to formulate the fundamental law of MTV, the Law of Value, according to which the magnitude of value of commodities regulates the proportions in which they exchange. In order to address this problem, I shall make use of techniques pertaining to contemporary logic and the theory of science to analyze

7

and reformulate MTV as given by Marx in the first two volumes of *C*. The leading thread shall be the requirement of producing a quantitative concept of value useful to formulate the Law of Value and to define the concept of surplus-value. In §1 I will introduce and discuss Marx's concept of value as it was presented by the author in the first chapter of *C*. In §2 I will describe an idealized fictitious market economy that I have labeled 'simple Marxian economy'; roughly speaking, this is the economy described by Marx himself mainly in the first two volumes of *C*. The task of producing a quantitative concept of value for this simple economy shall be tackled in §3.

Once we reach a clear cogent restatement of MTV, we will be in an advantageous position to formulate the foundational problems of the theory. We shall see throughout the exposition that all these foundational problems of MTV are clustered around the problems of providing a fairly general quantitative concept of value useful to give a correct general formulation of the Law of Value.

1.1 MARX'S CONCEPT OF VALUE

As it is well known, the starting point of Marx in *C* is the concept of commodity. A commodity is a use-value, a useful thing which is produced by a capitalist firm with the purpose of selling it in the market. Assuming that use-values are produced in different kinds and that there is a standard or unit of measurement for these objects, Marx proceeds to analyze the exchange process, in which commodities of different kinds exchange in the market in certain proportions. Introducing the concept of exchange-value, as "the quantitative relation, the proportion in which use-values of one kind exchange for use-values of another kind",[2] Marx proceeds to consider whether exchange-value is something merely relative and accidental, or whether, on the contrary, there is an intrinsic value, inherent to the commodity. From the fact that the exchange value of one kind of commodity is represented by specified amounts of goods of other kinds,

Marx infers that any of these two amounts must be mutually re-placeable or "of identical magnitude", and then claims:

> It follows from this that, firstly, the *valid* exchange-values [*gültigen Tauschwerte*] of a particular commodity express something equal, and secondly, exchange-value cannot be anything other than the mode of expression, the 'form of appearance', of a content distin-guishable from it.[3]

Considering the "equation" '1 quarter of corn = x cwt of iron', Marx asks: What does this equation signify?; and responds the following:

> It signifies that a common element of identical magnitude exists in two different things, in 1 quarter of corn and similarly in x cwt of iron. *Both are therefore equal to a third thing, which in itself is neither the one nor the other. Each of them, so far as it is exchange value, must therefore be reducible to this third thing.* [...] the exchange values of commodities *must* be reduced to a common element, of which they represent a greater or a lesser quantity.[4]

Since the exchange relation of commodities is characterized precisely by its abstraction from their use-values, and hence from the natural properties of the goods involved, Marx infers that this common element cannot be a natural property of them. Then, after seemingly reflecting about which properties are left, he claims that if we "disregard the use-value of commodities, only one property remains, that of being products of labor". In this way, through a sort of abstractive process, Marx introduces for the first time in *C* his concept of value as labor. It is a plausi-ble abductive reasoning by which he intends to establish that the labor required to produce the commodities is the only factor by which they are treated as equivalents in the exchange process, provided that the exchange-values of the commodities are *valid*. Methodologically, this procedure suggests that value is arrived at *a posteriori*, that value is somehow discovered in the exchange relation, that it is manifested in valid exchange-value.

After the former revelation has taken place, Marx goes on to consider the nature of value "independently of its form of appearance". This methodological turn indicates that Marx thinks that this object (that is, value) so discovered can be determined independently of its form of manifestation (which is valid exchange-value). This methodological decision —as we shall see— has enormous implications for the understanding of MTV, for the way Marx treats value along the remaining part of section 1, and throughout section 2, presupposes that the value of commodities —both the substance and magnitude of its value— not only can, but also *must* be determined independently of the market. This is apparent from the outset, since Marx is clearly bringing forth what he calls valid exchange-value. What could distinguish valid from non-valid exchange-value? It will be plain that, once we adopt the above mentioned methodological decision of severing the determination of value from the exchange process, the ground for the distinction between valid and non-valid exchange-value cannot be but the value of commodities *as determined independently of the market*: a valid exchange-value is one by which commodities exchange by their values, *where these values are determined solely in the sphere of production*. We will show that this is indeed so in *C*, and that it shall give rise to a very important contradiction between the first two volumes of *C* and the third, a contradiction pointed out in due time by Böhm-Bawerk in his "The Conclusion of the Marxian System" (1896), as well as to many other foundational difficulties that will appear later. Before considering all this, I would like to convey to the reader what Marx "really meant" by 'value' —at least in *C*— and to provide, out of Marx's writings, a new systematic formulation of his theory of value, a formulation which —being closer to our standards of clarity— will entitle us to dissect the foundational problems that it posed for posterity. We shall begin to consider these in the next chapter, starting with Böhm-Bawerk's criticisms.

Right after finishing his argument from the equivalence of exchangeable commodities, Marx asserts that "a use-value, or useful article, therefore, has value only because abstract human labor is objectified or materialized in it". This claim, again, sounds as if Marx had made an empirical discovery: he seems to have found that, as a matter of fact, commodities can exchange as values in certain proportions due exclusively to a "property" they possess, namely, that of being incarnations of objectified abstract human labor. From this point of view, commodities are nothing but

> merely congealed quantities of homogeneous human labor, i.e. of human labor-power expended without regard to the form of its expenditure. All these things now tell us is that human labor-power has been expended to produce them, human labor is accumulated in them. As crystals of this social substance, which is common to them all, they are values —commodity values [*Warenwerte*].[5]

We shall refer to this characterization of value as congealed quantities of homogeneous human labor as 'the first definition of value'. The first question that Marx rises in connection with value thus defined concerns the way in which it can be measured:

> How, then, is the magnitude of this value to be measured? By means of the quantity of the 'value-forming substance', the labor, contained in the article. This quantity is measured by its duration, and the labor-time is itself measured on the particular scale of hours, days, etc.[6]

Marx conceived the total labor-power of society as one homogeneous mass of human labor-power, composed of individual units of labor-power, claiming that each of these units

> is the same as any other, to the extent that has the character of a socially average unit of labor-power and acts as such, i.e. only needs, in order to produce a commodity, the labor-time which is necessary on average, or in others words is socially necessary. *Socially necessary labor-time is the labor-time required to produce any use-value under the conditions of production normal for a given society and with the average degree of skill and intensity of labor prevalent in that society* [...] *what exclusively determines the magnitude of the value of any article is therefore the amount of labor socially necessary, or the labor-time socially necessary for its production.*[7]

This characterization of the magnitude of value, as the amount of socially necessary labor-time required to produce the commodity, introduces in fact a second characterization of value, that we shall label 'the second definition of value'. On the other hand, the claim that each individual unit of labor-power is the same as any other, is one that needs justification. As Marx himself acknowledges a few paragraphs later, the different units of labor-power are not homogeneous:

> The totality of heterogeneous use-values or physical commodities reflects a totality of similarly heterogeneous forms of useful labor, which differ in order, genus, species and variety: in short, a social division of labor. This division of labor is a necessary condition for commodity production [...][8]

Since Marx insists that "the value of a commodity represents labor pure and simple, the expenditure of human labor in general", he must needs solve the problem posed by the heterogeneity of labor. In order to do so, Marx seems to perform an intellectual operation, pointing out that "if *we* leave aside the determinate quality of productive activity, and therefore the useful character of the labor, what remains is its quality of being an expenditure of human labor-power".[9] After this operation, Marx proceeds to introduce the concept of simple average labor, which takes as something given in any society, and goes on to assert that in experience we find that a reduction of more complex labor to simple labor is effected, since in fact a commodity which is the outcome of more complex labor is posited "through its value" as equal to the product of simple labor. It is not clear, unfortunately, what Marx could have meant by saying that complex labor is reduced to simple labor "through the value" of the corresponding produced commodities. He is trying to characterize value in terms of simple average labor, and then goes on to point out that the reduction of complex labor to this one is effected by value! It seems to me that Marx should have said here 'exchange-value' instead of just 'value'. Had he done so, the passage could be read as claiming that *the market*, that is *the exchange process*, is the social process —"that goes behind the backs of the producers"— that establishes "the various proportions in which different kinds of labor are reduced to simple labor as their unit of measurement".

Nevertheless, Marx does not say so, leaving in obscurity the inner workings of such process, because —I think— to say it would have been inconsistent with his declared purpose of considering the nature of value "independently of its form of appearance", that is of exchange-value. Rather, trying to be consequent with that purpose, Marx solves the problem by means of an act of faith, saying that

> In the interests of simplification, we shall henceforth view every form of labor-power directly as simple labor-power; by this we shall simply be saving *ourselves* the trouble of making the reduction. [...] Just as, in viewing the coat and the linen as values, *we* abstract from their different use-values, so, in the case of the labor represented by those values, do *we* disregard the difference between its useful forms, tailoring and weaving.[10]

Thus, Marx is in fact assuming that the reduction of complex to simple labor, or, more generally, the reduction of the different heterogeneous labor to a common unit of measurement, is something which could be done independently of the market; it is only for theoretical purposes that we save ourselves, i.e. the econometricians or the theoreticians, the trouble of making the reduction. In this form Marx finally arrives at his notion of homogeneous labor, which he also calls 'abstract labor', claiming that

> all labor is an expenditure of human labor-power, in the physiological sense, and it is in this quality of being equal, or abstract, human labor that it forms the value of commodities.[11]

Therefore, it is fairly clear that the impression Marx gives to the reader in the first two sections of *C* is that through valid exchange-value —which is the "form" of value even though someone may be in darkness concerning whether a particular exchange-value is valid or not— we become aware of the existence of value as homogeneous labor —which is the "substance" of value. The reader is also given the impression that in order to measure this substance in terms of socially necessary labor-time —which is the "magnitude" of value— we have to perform first the operation of reducing complex and heterogeneous labor to a common unit (simple average labor) and then to perform direct time measurements over all the production processes. In other words, Marx instills in these sections the methodological maxim that the magnitude of value is one that has to be determined independently of the market. If Marx would have claimed that this was not his intention, he should have granted at least that his exposition failed to convey what he really meant to say, being his second crucial section particularly obscure and sloppy.

In section 3, devoted to the different value-forms, Marx makes a claim that seems to be incompatible to what we saw he had said in the previous section. Even though Marx in fact assumed that the reduction of complex to simple labor, or, more generally, the reduction of the different heterogeneous labors to a common unit of measurement, is something which could be done independently of the market (say by a scientific observer endowed with the relevant required information), in section 3 Marx claims that this reduction is actually effected by the market:

> By equating, for example, the coat as a thing of value to the linen, we equate the labor embedded in the coat with the labor embedded in the linen. Now it is true that the tailoring which makes the coat is concrete labor of a different sort from the weaving that makes the linen. But *the act of equating tailoring with weaving reduces the former in fact to what is really equal in the two kinds of labor, to the characteristic they have in common of being human labor* [...] It is only the expression of equivalence between different sorts of commodities which brings to view the specific character of value-creating labor, *by actually reducing the different kinds of labor embedded in the different kinds of commodity to their common quality of being human labor in general.*[12]

If the reduction of the different labors is effected by the market, the doubt about the coherence of Marx account arises because in such a case it does not seem necessary to theoretically make that reduction: in such a case it is the *market* what saves ourselves the trouble of making the reduction; all we have to do as theoreticians or econometricians is to find out in the economic world just how the market made the reduction. This view brings important implications for the concept of value. We saw how Marx found the objective manifestation of value in exchange-value, presenting his unveiling of value through exchange-value as a sort of empirical discovery. Now, one thing is to say that value is discovered or manifested through exchange-value, and quite another to say that the value commodities have happens to be precisely the one manifested by any given system of exchange proportions. For instance, suppose that the price of all commodities is set with respect to gold. Then the labor embedded in all commodities is reduced in fact, by the market, to the labor embedded in gold. What this means is that we can take gold mining as the standard unit to compare all labors, and so all concrete labors can be expressed in terms of units of socially necessary mining time. As Marx puts it:

> The body of the commodity, which serves as the equivalent, always figures as the embodiment of abstract human labor, and is always the product of some specific useful and concrete labor. This concrete labor therefore becomes the expression of abstract human labor. [...] The equivalent form therefore possesses a second peculiarity: in it concrete labor becomes the form of manifestation of its opposite, abstract human labor. [...And] possesses the characteristic of being identical with other kinds of labor. [...] The equivalent form has a third peculiarity: private labor takes the form of its opposite, namely labor in its direct social form.[13]

More specifically and for the sake of illustration: if we say that one coat is worth two ounces of gold, that it takes one hour of mining to produce one ounce of gold and three hours of tailoring to produce one coat, then we would have to say that three hours of tailoring are equivalent to two hours of gold mining. If the market reduces all concrete labors in this way to mining time, then the door is open to *define* the measure of the magnitude of value of any good as the quantity of socially necessary mining-time to which it is equivalent. Clearly, this definition of the measure of value seems to involve an essential appeal to the actual exchange proportions in the market, making essentially dependent any measurement or determination of value on the proportions in which goods are actually exchanged in the market. One consequence of this is, of course, the Law of Value in its strongest form: the price of any bundle of goods is proportional to its value. In this case, the claim that commodities are exchanged by their values would seem to be no longer a daring synthetic proposition, but a consequence of the definition of 'value'. Clearly, all these theses are inconsistent with the idea that value determination is a business that has to be carried on without taking the market into account, independently of the exchange process.

Yet the idea that values have some autonomy with respect to the motions taking place in the market pops up again in section 3, just in the following passage:

> The equation 20 yards of linen = 1 coat, or 20 yards of linen are worth one coat, *presupposes* the presence in 1 coat of exactly as much of the substance of value as there is in 20 yards of linen, implies therefore that the quantities in which the two commodities are present *have cost the same amount of labor or the same quantity of labor-time*.[14]

Clearly, this idea —which is further confirmed by Marx's analyses of the influence of productivity on the relative expression of the magnitude of value— sends us back to the idea that somehow the substance of value does not depend on the exchange proportions; for otherwise it would not make sense the proviso that the former equation "presupposes" the presence of the same substance of value in both bundles of commodities. In the market-dependent view, the equation does not presuppose anything, it rather *establishes* the values. Moreover, after all this logical swinging, Marx seems to make his mind up and finally decides toward the end of section 3 to sustain the market-independent view, when he asserts the Law of Value in the following form:

> It becomes plain that it is not the exchange of commodities which regulates the magnitude of their values, but rather the reverse, the magnitude of the value of commodities which regulates the proportions in which they exchange.[15]

Clearly, there appears to be certain inconsistency in Marx treatment along sections 1, 2 and 3 of the first chapter of *C.* There appears to be a contradiction between the idea that value is something determined in the sphere of production, measured in terms of the socially necessary labor-time required for the production of the different commodities, by means of a reduction of heterogeneous to simple homogeneous labor, which would have to be made in a way Marx never explains, and the idea that value is the outcome of exchange, which actually reduces all concrete private labors to a common unit. This contradiction is indeed the first foundational problem posed by MTV: depending on which horn of the dilemma we choose to adopt, we can reach two very different versions of MTV. It turns out that most interpreters and critics of Marx have chosen the one that Marx seems to support most, namely the market-independent view. This is the interpretation followed by Böhm-Bawerk and more modern (and less destructive) readers of Marx, although at least one of his interpreters, Isaak Illich Rubin, tried to follow the market-dependent view. I submit that it does not make much sense for foundational work in MTV to decide which was, in the end, Marx's "ultimate position" on the matter. Whatever it was, he took it with him to the grave. My methodological stance toward this dilemma is the following, to wit, that the final decision between these two interpretations (if it is true that they cannot be made compatible to some extent) must be taken on the basis of which one solves more logical problems of MTV, keeping in mind that the leading thread of the present foundational work is the requirement of producing a quantitative concept of value useful to define the concept of surplus-value and establish the Law of Value.

As a methodological decision, I shall adopt here the market-independent view of value as the core of the prototype of MTV, and proceed accordingly to provide a logical reconstruction of this theory based upon that view. Trying to stick as much as possible to the description of the valorization process as given by

Marx in *C*, I will try to make explicit, in a way which is canonical according to current logical standards, the minimum structure or set of axioms which guarantee the existence of market-independent values for an economy of the sort described by Marx in his formulation of the theory of value.

1.2 DESCRIPTION OF THE PROTOTYPE

Throughout *C*, but mainly in the first two volumes, Marx describes a fictitious idealized market economy —which I shall label 'simple Marxian economy'— in order to discuss the laws that explain the functioning of the capitalist mode of production. This economy is characterized by a set of *m* formally independent producers, each one producing a particular kind of good, so that *m* different kinds of goods are produced altogether in the economy. The theory represents the workings of the economy during one certain interval of time (say one year). At the beginning of the year all factories (one for each producer) are endowed with brand new equipment, all the prime materials they will need, and hired personnel representing simple homogeneous labor-power. The capital goods in the factories have the property that can only be operated at a certain rythm, so that the same time of use always yields the same amounts of goods. Also, the workers never get too tired, and so their rythm of production is constant all of the time. Hence, if the output of production were increased or decreased by any fixed amount (if the owners decided to produce, for instance, twice as many goods in the same amount of time) then the amount of workers and equipment would have to be increased or decreased by the same amount (these technologies are described in the economic literature as yielding 'constant returns to scale'). Each factory operates every day during the same number of hours, the workers are never absent and there never are strikes, power failures or any problem preventing the normal operation of the factories. Moreover, all products have the same period of production, say the whole year, and the period of rotation of capital goods is such that at the end of the period, at the end of the year, they all are entirely worn out and have to be replaced anew. Hence, the last day of the year appear all of a sudden all final net outputs (one unit for each factory) together with the inputs produced along the period for replacement purposes, all original prime materials have been entirely consumed (as planned) and all the original machines simultaneously break down, beyond any possibility of repair, exhalating their last breath after an exemplarly productive life (thus, in contemporary terminology, these processes are of the point-input-point-output type).

Other traits characterizing the former, rather regular world, are that there is no choice of techniques in the economy, i.e. only one production procedure is used for each kind of commodity. There is no joint production, i.e. each factory produces only one kind of good as output. Also, every kind of good is produced under oligopolic conditions, i.e. each kind is produced by one producer only, although different kinds are produced by different producers. Supposing that the reduction of heterogeneous labor to a common standard can be effected as Marx suggested, all labors in the different factories are taken as simple homogeneous labor and so there are no heterogeneous concrete labors. It is also assumed that every industry requires a positive amount of labor. In order to reproduce the labor-power expended in one hour of work, it is necessary the consumption of certain amounts of wage goods which taken together will constitute the consumption basket of the working class. We shall suppose that the wage of the workers is just enough to acquire this basket. We suppose also that the economy is closed, which means that all kinds of means of production are produced in the economy. The wage and capital goods factories are semiproductive, i.e. they are at least able to reproduce themselves, in the sense that they consume no more than they produce. Finally, all the wage and capital goods factories are interconnected, i.e. there is no independent subgroup of capital or wage good industries, that is to say, no subgroup of factories producing capital or wage goods, which does not require to employ outputs produced by another subgroup.

The previous economic fiction was not described by Marx in a complete manner, but was sketched by him along the first two volumes of *C* (although in the second volume he makes room for capital goods having different rotation periods). The additional traits of the economy, those not explicitly given by Marx, were introduced to guarantee the existence of a unique quantity of unskilled labor associated to each unit of good produced in the economy, i.e. to guarantee the existence of a unique numerical labor-value for each unit of commodity produced within the system. This fiction can be mathematically modeled, a necessary step in order to prove formally the existence of such numerical magnitudes. The mathematical modeling of this economy and of some others that will be introduced later, as well as the formulation of the foundational problems of MTV, requires the introduction of some basic notation at this point.

1.3 MATHEMATICAL MODELING OF THE PROTOTYPE

I ask the reader to grant the adoption now of certain notational conventions. In *C* Marx distinguished in the labor process three main constituents: (1) the labor-power, (2) the instruments of labor, and (3) the objects of labor.[16] The instruments and objects of labor are called generically by Marx 'means of production', but usually we will call them 'inputs', following a widespread current practice. These constituents, as exemplified in a particular really existing production process (say in one journey of a Volkswagen factory), can certainly be *measured* in terms of suitable units. In the case of labor-power, suppose that there are in the economy n different trades, i.e. n different kinds of concrete labors (weaver, tailor, mason, carpenter, civil engineer, etcetera). Every expenditure of any of these kinds is to be measured —as Marx suggested— in terms of labor-time: x_1 hours of weaving, x_2 hours of tayloring, and so on, where x_i ($1 \leq i \leq n$) is a nonnegative real number. Thus, the labor-power actually applied in the process can be represented mathematically as an n-dimensional vector $\mathbf{x}$:

$$\mathbf{x} = [x_1 \cdots x_n].$$

If m kinds of goods are produced in the economy, we can measure these goods in terms of fixed physical units of measurement; for instance, 20 yards of linen, 1 coat, half a ton of iron, and so on. We shall adopt from now on the convention that the first k components of vector

$$\underline{\mathbf{x}} = [\underline{x}_1 \cdots \underline{x}_k \underline{x}_{k+1} \cdots \underline{x}_m]$$

represent amounts of capital goods, whereas the remaining $m - k$ represent amounts of luxury or wage consumption goods; we assume, moreover, that each position of such vectors is conventionally associated with one kind of good. This means that goods belonging to the kinds represented by positions $k + 1$ to m are never used as production means in any production process of the economy, whereas those represented by the first entries, 1 to k, appear in such processes. We adopt the convention that vectors of the form $\underline{\mathbf{x}}$ represent inputs of some identifiable production process. Obviously, this entails that in vectors $\underline{\mathbf{x}}$ the last $m - k$ entries are zero. In a similar way, vectors of the form

$$\overline{\mathbf{x}} = [\overline{x}_1 \cdots \overline{x}_n]$$

will represent outputs of the corresponding production process; in these vectors, any entry can be positive, since the economy may produce both capital as well as wage and luxury goods. Usually, we shall gather vectors $\mathbf{x}$, $\underline{\mathbf{x}}$ and $\overline{\mathbf{x}}$ into a single vector $\tilde{\mathbf{x}}$, in the following way:

$$\tilde{\mathbf{x}} = [\mathbf{x}, \underline{\mathbf{x}}, \overline{\mathbf{x}}].$$

Hence, vectors of this form represent production processes. The net output of production process $\tilde{\mathbf{x}}$ we shall represent by means of the symbol $\hat{\mathbf{x}}$, which is defined as the difference $\overline{\mathbf{x}} - \underline{\mathbf{x}}$. The generic entries of vectors $\mathbf{x}$, $\underline{\mathbf{x}}$, $\overline{\mathbf{x}}$ and $\hat{\mathbf{x}}$ will be written in a natural way as x_i, $\underline{x}_i$, $\overline{x}_i$ and $\hat{x}_i$, respectively, with i running in each case over the appropriate set of indexes. In particular, if these components are zero, the vectors will be written as $\mathbf{0}$, $\underline{\mathbf{0}}$, $\overline{\mathbf{0}}$ and $\hat{\mathbf{0}}$ respectively; whereas the whole production processes will be written as $\widetilde{\mathbf{0}}$. Given any two vectors of the same dimension, say $\mathbf{w} = [w_1 \cdots w_n]$ and $\mathbf{x} = [x_1 \cdots x_n]$, their inner product

$$\sum_{i=1}^{n} w_i x_i$$

shall be denoted by the juxtaposition $\mathbf{wx}$ of vectors $\mathbf{w}$ and $\mathbf{x}$.

The symbol $\leq$ among numerals shall mean that the first number is less than or equal to the second; among vector symbols it means that some entry of the first vector is strictly less than the corresponding entry of the second, and that no entry of the second is strictly less than the corresponding entry of the first; $\leqq$ is written only among vector symbols and it means that the first vector is equal to the second or that the relation expressed by $\leq$ holds. The symbol $<$ among numerals means that the first number is strictly less than the second; among vector symbols it means that all entries of the first vector are strictly less than their corresponding entries in the second. The symbols $\geq$, $\geqq$ and $>$ express, respectively, the converses of the relations expressed by $\leq$, $\leqq$ and $<$.

After this brief notational digression, let us return to our previous concern, the mathematical modeling of our simple Marxian economy. In this economy there are m producers, each one produces only one kind of good (no joint production), no two producers produce the same kind of good (oligopoly) and there is no choice of tecniques, so that every good is produced by one producer using one fixed tecnology. Hence, at the beginning of the year producer i ($1 \leq i \leq m$) has production means represented by vector $\underline{\mathbf{x}}_i = [\underline{x}_{1i} \cdots \underline{x}_{mi}]$, which are entirely consumed along the year in order to produce one unit of good of kind i, represented by vector $\overline{\mathbf{x}}_i = [\overline{x}_{1i} \cdots \overline{x}_{mi}]$, where $\overline{x}_{ij} = 0$ for every $j \neq i$, and $\overline{x}_{ii} = 1$. Since labor is unskilled, there are no heterogeneous concrete labors and so the vector of labor inputs $\mathbf{x}_i$ of process i is unidimensional, i.e. it is a scalar, that shall be represented by x_i. Thus, following our previous conventions, the industries producing capital goods are represented by vectors

$\tilde{\mathbf{x}}_1, \tilde{\mathbf{x}}_2, \ldots, \tilde{\mathbf{x}}_k$; the industries producing wage and luxury goods by $\tilde{\mathbf{x}}_{k+1}, \tilde{\mathbf{x}}_{k+2}, \ldots, \tilde{\mathbf{x}}_m$. Now, let $\underline{\mathbf{x}}_1^T, \ldots, \underline{\mathbf{x}}_k^T$ be the transposes of vectors $\underline{\mathbf{x}}_1, \ldots, \underline{\mathbf{x}}_k$ after their last $m - k$ components (which happen to be all zero) have been dropped. Analogously, let the vectors $\underline{\mathbf{x}}_{k+1}^T, \ldots, \underline{\mathbf{x}}_m^T$ be the transposes of vectors $\underline{\mathbf{x}}_{k+1}, \ldots, \underline{\mathbf{x}}_m$ after their last $m - k$ components (which also happen to be all zero) have been eliminated. Then the matrix of capital good industries, $\mathbf{A}_I$, is defined by

$$\mathbf{A}_I = [\underline{\mathbf{x}}_1^T \cdots \underline{\mathbf{x}}_k^T],$$

whereas the matrix $\mathbf{A}_{II}$ of consumption goods industries is defined by

$$\mathbf{A}_{II} = [\underline{\mathbf{x}}_{k+1}^T \cdots \underline{\mathbf{x}}_m^T].$$

Clearly, $\mathbf{A}_I$ is a square $k \times k$ matrix, while $\mathbf{A}_{II}$ is rectangular $k \times (m - k)$.

The labor inputs of processes $\tilde{\mathbf{x}}_1$ to $\tilde{\mathbf{x}}_m$ are collected in the matrices

$$\mathbf{L}_I = [x_1 \cdots x_k] \quad \text{and} \quad \mathbf{L}_{II} = [x_{k+1} \cdots x_m].$$

Notice that since every industry requires a positive amount of labor, $\mathbf{L}_I$ and $\mathbf{L}_{II}$ are both positive.

Now, according to Marx's first definition of value, as crystals of homogeneous human labor commodities are values. Let λ_i be the amount of value congealed in one unit of commodity of kind i $(i = 1, \ldots, k)$. Since, according to Marx, the value of a product is composed by the value of the production means expended in its production process, plus the amount of direct live labor invested in the same, the magnitude of value of one unit of capital good i must satisfy equation

(1) $$\lambda_i = \underline{x}_{1i}\lambda_1 + \cdots + \underline{x}_{ki}\lambda_k + x_i.$$

Analogously, the magnitude λ_j $(j = k + 1, \ldots, m)$ of value of a unit of non-capital good of kind j, satisfies equation

(2) $$\lambda_j = \underline{x}_{1j}\lambda_1 + \cdots + \underline{x}_{kj}\lambda_k + x_j.$$

(Notice that we are writing the components of a vector $\underline{\mathbf{x}}_l$ as $\underline{x}_{1l}$, $\ldots$, $\underline{x}_{kl}$ for $l = 1, \ldots, m$). All formulas of type (1) and (2) can be put in concise form as

$$(3) \qquad \Lambda_{\mathrm{I}} = \Lambda_{\mathrm{I}} \mathbf{A}_{\mathrm{I}} + \mathbf{L}_{\mathrm{I}} \qquad \text{and} \qquad \Lambda_{\mathrm{II}} = \Lambda_{\mathrm{I}} \mathbf{A}_{\mathrm{II}} + \mathbf{L}_{\mathrm{II}}.$$

provided that

$$\Lambda_{\mathrm{I}} = [\lambda_1 \cdots \lambda_k] \quad \text{and} \quad \Lambda_{\mathrm{II}} = [\lambda_{k+1} \cdots \lambda_m].$$

These are the equations for value implied by the first definition of value.

According to the second definition, only the amount of socially necessary labor-time required for the production of a commodity determines its magnitude of value. In order to compute this magnitude, let us consider the total amounts of capital goods $y_{1i}, \ldots, y_{ki}$ which are required for the production of a unit of capital good i, after taking into account all the repercussions. In order to determine these amounts, the following equation must be solved:

$$(4) \qquad \qquad \mathbf{A}_{\mathrm{I}} \mathbf{y}_i + \overline{\mathbf{x}}_i^{\mathrm{T}} = \mathbf{y}_i,$$

where $\overline{\mathbf{x}}_i^{\mathrm{T}}$ is the trasepose of output vector $\overline{\mathbf{x}}_i$, after its last $l - k$ coordinates have been dropped, and

$$\mathbf{y}_i = [y_{1i} \cdots y_{ki}]^{\mathrm{T}}.$$

Assuming that equation (4) has a unique solution $\mathbf{y}_i$, the amount of socially necessary labor-time μ_i required to produce one unit of capital good i is given by

$$(5) \qquad \qquad \mu_i = \mathbf{L}_{\mathrm{I}} \mathbf{y}_i.$$

On the other hand, since capital goods $\underline{x}_{1j}, \ldots, \underline{x}_{kj}$ (which are the entries of $\underline{\mathbf{x}}_j^{\mathrm{T}}$) are required for the production of one unit of consumption good j $(j = k + 1, \ldots, m)$, for replacement purposes goods must be produced in the amounts $y_{1j}, \ldots, y_{kj}$ determined by equation

$$(6) \qquad\qquad \mathbf{A_I y_j} + \mathbf{\underline{x}}_j^T = \mathbf{y_j},$$

$(\mathbf{y_j} = [y_{1j} \cdots y_{kj}]^T)$ in order to provide the wage or luxury goods industry with net outputs of capital goods in the amounts which are just enough for the replacement.

Given that in the production of the capital goods $\mathbf{L_I y_j}$ units of labor-time are expended, and the production of the consumption good j adds x_j units of direct labor to its product, the total socially necessary labor-time required to produce one unit of wage or luxury good j $(k + 1 \leq j \leq m)$ is given by

$$\mu_j = \mathbf{L_I y_j} + x_j.$$

Hence, if we set $\mathbf{M_I} = [\mu_1 \cdots \mu_k]$, $\mathbf{M_{II}} = [\mu_{k+1} \cdots \mu_m]$, $\mathbf{Y_I} = [\mathbf{y_1} \cdots \mathbf{y_k}]$ and $\mathbf{Y_{II}} = [\mathbf{y_{k+1}} \cdots \mathbf{y_m}]$, we get the equations

$$\mathbf{A_I Y_I} + \mathbf{I} = \mathbf{Y_I}, \qquad \mathbf{A_I Y_{II}} + \mathbf{A_{II}} = \mathbf{Y_{II}}$$

and

$$(7) \qquad\qquad \mathbf{M_I} = \mathbf{L_I Y_I}, \qquad \mathbf{M_{II}} = \mathbf{L_I Y_{II}} + \mathbf{L_{II}}.$$

Notice that $\mathbf{I} = [\mathbf{\bar{x}}_1^T \cdots \mathbf{\bar{x}}_k^T]$ is the identity matrix.

The former are the equations for value according to Marx's second definition. The first question that arises is whether the value equations (3) and (7) have unique positive solutions and, if the answer to this question is affirmative, the second question is whether values as determined according to the first definition of value coincide with values as determined according to the second one, i.e. whether $\Lambda_I = \mathbf{M_I}$ and $\Lambda_{II} = \mathbf{M_{II}}$. The answer is affirmative in both cases, but in order to prove it, it will be necessary to express in mathematical terms the semiproductivity and interconnectedness of the wage and capital goods industries of our simple economy.

In order to give mathematical expression to the reproducibility or semiproductivity and the interconnectedness of the wage and capital goods industries of a simple economy, let us represent first the consumption basket of the working class, namely as the column vector $\mathbf{b}$:

$$\mathbf{b} = \begin{bmatrix} b_{k+1} \\ \vdots \\ b_l \end{bmatrix}.$$

The amounts of goods $b_{k+1}, \ldots, b_l$ ($l \leq m$) are those that are necessary to reproduce the labor-power expended in one hour of labor. It is also convenient to introduce the matrix

$$\mathbf{A} = \begin{bmatrix} \underline{x}_{11} & \cdots & \underline{x}_{1k} & \underline{x}_{1k+1} & \cdots & \underline{x}_{1l} \\ \vdots & & \vdots & \vdots & & \vdots \\ \underline{x}_{k1} & \cdots & \underline{x}_{kk} & \underline{x}_{kk+1} & \cdots & \underline{x}_{kl} \\ 0_{k+11} & \cdots & 0_{k+1k} & 0_{k+1k+1} & \cdots & 0_{k+1l} \\ \vdots & & \vdots & \vdots & & \vdots \\ 0_{l1} & \cdots & 0_{lk} & 0_{lk+1} & \cdots & 0_{ll} \end{bmatrix}$$

where $\underline{x}_{1j}, \ldots \underline{x}_{kj}$ ($1 \leq j \leq l$) are the inputs of process $\tilde{\mathbf{x}}_j$, the other entries of the matrix being 0; processes $\tilde{\mathbf{x}}_1, \ldots \tilde{\mathbf{x}}_k$ are (as we had said) those of the capital good industries, and processes $\tilde{\mathbf{x}}_{k+1}, \ldots \tilde{\mathbf{x}}_l$ are those of the industries producing wage goods, i.e. goods of the types that constitute vector $\mathbf{b}$. Consider now the $l \times l$ matrices

$$\mathbf{B} = \begin{bmatrix} 0_1 & \cdots & 0_1 \\ \vdots & & \vdots \\ 0_k & \cdots & 0_k \\ b_{k+1} & \cdots & b_{k+1} \\ \vdots & & \vdots \\ b_l & \cdots & b_l \end{bmatrix}$$

and

$$\mathbf{L} = \begin{bmatrix} x_1 & \cdots & 0 \\ \vdots & & \vdots \\ 0 & \cdots & x_l \end{bmatrix}.$$

B contains only zeros above the $k + 1$th row and the last components of the columns are nothing but those of the consumption vector **b**. All off-diagonal entries of **L** are zero and the diagonal element of row i $(i = 1, \ldots, l)$ is the labor input of process producing wage or capital good of kind i. These matrices are useful to construct the product

$$\mathbf{BL} = \begin{bmatrix} 0_1 & \cdots & 0_1 \\ \vdots & & \vdots \\ 0_k & \cdots & 0_k \\ b_{k+1}x_1 & \cdots & b_{k+1}x_l \\ \vdots & & \vdots \\ b_l x_1 & \cdots & b_l x_l \end{bmatrix}.$$

It can be seen that the ith column of this matrix $(i = 1, \ldots, l)$ represents the amounts of wage goods $b_{k+1}x_i, \ldots, b_l x_i$ required to reproduce the labor-power expended in processes $\tilde{\mathbf{x}}_i$. Therefore, matrix

$$\mathbf{C} = \mathbf{A} + \mathbf{BL}$$

$$= \begin{bmatrix} \underline{x}_{11} & \cdots & \underline{x}_{1l} \\ \vdots & & \vdots \\ \underline{x}_{k1} & \cdots & \underline{x}_{kl} \\ b_{k+1}x_1 & \cdots & b_{k+1}x_l \\ \vdots & & \vdots \\ b_l x_1 & \cdots & b_l x_l \end{bmatrix}$$

$$= \begin{bmatrix} c_{11} & \cdots & c_{1l} \\ \vdots & & \vdots \\ c_{k1} & \cdots & c_{kl} \\ c_{k+11} & \cdots & c_{k+1l} \\ \vdots & & \vdots \\ c_{l1} & \cdots & c_{ll} \end{bmatrix}$$

represents the amounts of production means and wage goods required to operate the processes producing capital and wage goods. Now we are in position to express in mathematical terms the reproducibility and interconnectedness of the wage and capital goods industries. We say that the capital and wage industries are *semiproductive* or *reproducible* iff there is a positive (column) vector $\mathbf{y}$ such that $\mathbf{C}\mathbf{y} \leqq \mathbf{y}$. Also, we say that the wage and capital goods industries are *interconnected* iff the matrix $\mathbf{C}$ is indecomposable, i.e. if it is not the case that $c_{ij} = 0$ for all indices j belonging to some proper nonempty subset J of $\{1, \cdots, l\}$ and indices i not belonging to the same subset. What this means is that industries represented by the indexes in J do not require as inputs means of production produced by the industries not represented by indexes in J. This implies that if the matrix $\mathbf{C}$ is indecomposable then any increase in the final output of any capital or wage goods industry necessarily increases the requirement of inputs for every other such industry.

Hence, the assumptions that the capital and wage goods industries in the economy are both semiproductive and interconnected are expressed by saying that the matrix $\mathbf{C}$ is semiproductive and indecomposable. As a matter of fact, the reproducibility and indecomposability of $\mathbf{C}$ imply that $\mathbf{A_I}$, the matrix of coefficients of capital goods industries, is also indecomposable and —what is stronger than mere semiproductivity— also quasiproductive, i.e. there is a positive vector $\mathbf{z}$ such that $\mathbf{A_I}\mathbf{z} \leq \mathbf{z}$. This can be seen as follows. Since $\mathbf{C}$ is semiproductive, $\mathbf{C}\mathbf{y} \leqq \mathbf{y}$ for some positive vector $\mathbf{y}$. Drop the last $l - k$ components of $\mathbf{y}$, obtaining a vector $\mathbf{z}$ of dimension k. Then, given that $\mathbf{A_I}$ is the left upper corner of $\mathbf{C}$ and wage good industries require positive inputs from the capital good industries (because $\mathbf{C}$ is indecomposable), it is easy to see that $\mathbf{A_I}\mathbf{z} \leq \mathbf{z}$; in other words, some of the first k components of $\mathbf{C}\mathbf{y}$ are strictly greater than the corresponding components of $\mathbf{A_I}\mathbf{z}$. That $\mathbf{A_I}$ is indecomposable follows from the fact that any increase in the final output of any capital or wage industry, in particular of any capital good industry, necessarily increases the requirement of inputs for all the wage and capital goods industries and so, in particular, for all of the second. In order to establish many of the results we are concerned with, it will be necessary to prove the following

LEMMA: *Let $\mathbf{E}$ be a semipositive $\kappa \times \kappa$ square matrix that is quasiproductive and indecomposable. Then $(\mathbf{I} - \mathbf{E})^{-1}$, where $\mathbf{I}$ is the identity matrix, is a positive matrix.*

Proof: Since $\mathbf{E}$ is quasiproductive, there is a $\mathbf{y} > \mathbf{0}$ such that $\mathbf{Ey} \leq \mathbf{y}$. We are going to show that $\mathbf{I} - \mathbf{E}$ possesses a semidominant diagonal, i.e. that there exist positive $y_1, \ldots, y_\kappa$ such that

$$y_i|1 - e_{ii}| \geq \sum_{j \neq i} y_j| - e_{ij}| \qquad (i = 1, \ldots, \kappa)$$

with at least one strict inequality.

In fact, for every $i = 1, \ldots, \kappa$, $0 < e_{i1}y_1 + \cdots + e_{i\kappa}y_\kappa \leq y_i$, where this inequality is strict for at least one i, and so

$$y_i - x_{ii}y_i \geq \sum_{j \neq i} x_{ij}y_j \geq 0.$$

Hence,

$$y_i|1 - x_{ii}| \geq \sum_{j \neq i} y_j| - x_{ij}| \geq 0$$

with at least one strict inequality.

Since $\mathbf{E}$ is indecomposable, so is $\mathbf{I} - \mathbf{E}$, and so it follows that $\mathbf{I} - \mathbf{E}$ is nonsingular. Moreover, since all off-diagonal elements of $\mathbf{I} - \mathbf{E}$ are nonpositive, it follows that $(\mathbf{I} - \mathbf{E})^{-1} > \mathbf{0}$.[17] □

We are now in position to establish the first main result of this chapter. This is

THEOREM 1: *In a simple Marxian economy, there exist unique systems of positive values Λ_I and Λ_{II} in the sense of the first definition of value. Also, there exist unique systems of positive values $\mathbf{M}_I$ and $\mathbf{M}_{II}$ in the sense of the second definition of value.*

Proof: Since $\mathbf{A}_I$ is quasiproductive and indecomposable, by the Lemma $(\mathbf{I} - \mathbf{A}_I)^{-1}$ exists and is positive. Hence, setting $\Lambda_I = \mathbf{L}_I(\mathbf{I} - \mathbf{A}_I)^{-1}$, $\mathbf{Y}_I = (\mathbf{I} - \mathbf{A}_I)^{-1}$ and $\mathbf{Y}_{II} = (\mathbf{I} - \mathbf{A}_I)^{-1}\mathbf{A}_{II}$, we see at once that the matrices Λ_I, Λ_{II}, $\mathbf{M}_I$ and $\mathbf{M}_{II}$ are unique and positive. □

It can be shown that, in fact, in a simple Marxian economy the first definition of value is equivalent to the second. The first one in having noticed that value can be characterized in a dual way was Michio Morishima (1973). He was also the first one in bringing to the light the hidden assumptions required to prove both the previous theorem and the next one.

THEOREM 2: *In a simple Marxian economy, values as determined according to the first definition coincide with values as determined according to the second one. That is to say, equations $\Lambda_I = M_I$ and $\Lambda_{II} = M_{II}$ hold.*

Proof: It suffices to see that the following chain of implications holds:

$$\Lambda_I = \Lambda_I A_I + L_I \Rightarrow \Lambda_I Y_I = \Lambda_I A_I Y_I + L_I Y_I$$
$$\Rightarrow \Lambda_I (Y_I - A_I Y_I) = L_I Y_I$$
$$\Rightarrow \Lambda_I I = L_I Y_I$$
$$\Rightarrow \Lambda_I = M_I.$$

This establishes the first identity. For the second one, using the fact that $\Lambda_{II} = \Lambda_I A_{II} + L_{II}$, we have:

$$\Lambda_I = \Lambda_I A_I + L_I \Rightarrow \Lambda_{II} + \Lambda_I Y_{II} = (\Lambda_I A_{II} + L_{II})$$
$$+ (\Lambda_I A_I Y_{II} + L_I Y_{II})$$
$$\Rightarrow \Lambda_{II} + \Lambda_I Y_{II} - \Lambda_I (A_I Y_{II} + A_{II})$$
$$= L_I Y_{II} + L_{II}$$
$$\Rightarrow \Lambda_{II} + \Lambda_I Y_{II} - \Lambda_I Y_{II} = M_{II}$$
$$\Rightarrow \Lambda_{II} = M_{II}. \ \square$$

Thus, it turns out that in a simple economy the quantitative concept of value is well defined. I said that one of the leading threads of our research was going to be the production of a quantitative concept of value useful to define the concept of surplus-value. Hence, we must show that the just defined concept of value does permit to define that crucial notion. I shall conclude the present chapter by introducing the concepts of system of hourly wages and system of prices, in order to prove the Fundamental Marxian Theorem (FMT) for our simple economy. This theorem asserts that the rate of profit in terms of money of any process is positive iff the rate of exploitation, in terms of value, is also positive. The FMT is important not merely as a "denunciation" of capitalism (as if exploitation were by itself the cause of evil in the capitalist societies) but first and foremost as the establishment of a condition that prices and wages in a market economy must satisfy in order for the economy to be feasible and able to reproduce itself. In other words, exploitation is a scientific concept required to formulate an important constraint on prices in a market economy.

In order to introduce the argument, let us adopt some additional notational conventions. Let us suppose that any unit of each kind i $(i = 1, \ldots, m)$ of commodities in the economy has a single price p_i. Then the prices of all commodities can be put together in a positive vector $[p_1 \cdots p_m]$. Nevertheless, it will be convenient, for the purposes of treating the case of the simple economy, to split this vector into two: the vector $\mathbf{p} = [p_1 \cdots p_l]$ $(l \leq m)$ of wage or capital goods, and the vector $\mathbf{p}' = [p_{l+1} \cdots p_m]$ of luxury goods.

Let us denote the hourly wage paid to a worker belonging to trade i $(i = 1, \ldots, n)$ as w_i. Then we can write the wage system as a vector $[w_1 \cdots w_n]$ and so the total amount of salaries paid in process $\tilde{\mathbf{x}} = [\mathbf{x}, \underline{\mathbf{x}}, \overline{\mathbf{x}}]$ is the inner product $\mathbf{wx}$. In order to discuss the particular case of the simple economy, it will be convenient to denote with $\mathbf{w}$ the initial segment of vector $[w_1 \cdots w_n]$ containing the hourly wages of workers of capital and wage good industries $1, \ldots, l$, and with $\mathbf{w}'$ the remaining segment. Since labor in our simple economy has been taken as simple and homogeneous, the components of both vectors are all equal to the scalar w, denoting in each case the hourly wage of simple homogeneous labor. The assumption that workers do not save, i.e. that the wage of the workers is just enough to acquire the consumption basket $\mathbf{b}$, implies that $\mathbf{pb} = \mathbf{w}$.

In a competitive market economy, no firm can survive unless it obtains profits by selling its product in the market. A necessary condition for this is that the production price be less than the selling price of the product. This can be expressed in general by means of the formulas

(8) $\mathbf{p}C < \mathbf{p}$

for wage and capital goods, and

(9) $\mathbf{p}C' < \mathbf{p}'$

for luxury goods, where

$$
C' = \begin{bmatrix}
\underline{x}_{1l+1} & \cdots & \underline{x}_{1m} \\
\vdots & & \vdots \\
\underline{x}_{kl+1} & \cdots & \underline{x}_{km} \\
b_{k+1}x_{l+1} & \cdots & b_{k+1m}x_m \\
\vdots & & \vdots \\
b_l x_{l+1} & \cdots & b_l x_m
\end{bmatrix}.
$$

Depending on the price and the wage systems, the profit yielded by the same production processes may be large or small. A measure of the size of the profit yielded by each unit of money invested in the process is given by the profit rate of the process. The profit rate of process $\tilde{\mathbf{x}}_i$ under price system $[\mathbf{p}, \mathbf{p}']$, $\pi(\tilde{\mathbf{x}}_i)$, is the ratio of the benefit to the production cost:

$$
\pi(\tilde{\mathbf{x}}_i) = \frac{\mathbf{p}\hat{\mathbf{x}}_i - wx_i}{wx_i + \mathbf{p}\underline{\mathbf{x}}_i} \qquad i = 1, \ldots, m.
$$

Clearly, when we multiply the profit rate of process $\tilde{\mathbf{x}}_i$ by its production costs we obtain the net profit yielded by that process.

The only reason why a capitalist invests in order to produce some kind of use-value, is the desire for profits. On the other hand, the reproduction of the workers requires, for each hour of labor, the minimum consumption basket **b** of wage goods, which are necessary to reproduce one hour of homogeneous labor-power. Hence, there are two basic constraints that a system of wages and prices must satisfy in order to set the industries into motion. The first is that the system must guarantee a profit rate greater than one for all industries; the second is that the salary obtained by any worker for each hour of labor be sufficient to buy the consumption basket **b**, i.e. $wx_i = \mathbf{pb}x_i$ for every $i = 1, \ldots, m$. It turns out that in our simple economy prices and salaries can be established that satisfy these constraints. I will show that this is true, first, with respect to the capital and wage industries; it will follow easily that it holds also for the luxury goods industries. As a matter of fact, an even stronger result can be obtained. We shall see that it is possible to find a unique profit rate greater that one and a (unique up to multiplication by a positive scalar) price vector for all production processes in the economy. Before this, we will prove that a price system **p** for wage and capital goods can be found that at least permits to operate the industries of these goods without any losses (even though perhaps also without any profits).

THEOREM 3:   *There is a unique nonnegative profit rate r to which there correspond price systems* **p** *and* **p'** *such that*

$$\mathbf{p} = (1 + r)\mathbf{pC} \qquad and \qquad \mathbf{p}' = (1 + r)\mathbf{pC}'.$$

Moreover, these price systems are unique up to scale transformations.

Proof: Since the matrix **C** is indecomposable, by the Perron-Frobenius theorem there is a unique positive real eigenvalue γ (the Frobenius root) to which there corresponds a unique (up to multiplication by a positive scalar) positive left eigenvector **p**:

$$\mathbf{pC} = \mathbf{p}\gamma.$$

Let

$$r = \frac{1 - \gamma}{\gamma}.$$

Since **C** is semiproductive, there is a positive vector **y** such that $\mathbf{Cy} \leqq \mathbf{y}$. Hence, $\mathbf{pCy} \leq \mathbf{py}$ and so $\gamma \leq 1$. It follows that r is a uniquely determined nonnegative number and that $1 + r = \gamma^{-1}$. Therefore,

$$\mathbf{p} = (1 + r)\mathbf{pC}.$$

We just set now the prices of luxury goods as required:

$$\mathbf{p}' = (1 + r)\mathbf{p}\mathbf{C}'. \ \square$$

The number r shall be called also the *equilibrium profit rate*. Any price systems $\mathbf{p}$ and $\mathbf{p}'$ satisfying equations

$$\mathbf{p} = (1 + r)\mathbf{p}\mathbf{C} \qquad \text{and} \qquad \mathbf{p}' = (1 + r)\mathbf{p}\mathbf{C}'$$

will be called *equilibrium price systems*. Clearly, if the equilibrium profit rate r is greater than zero, then the price systems corresponding to it satisfy equations (8) and (9).

The existence of a positive uniform profit rate is not necessary, but it can be shown that a necessary condition for its existence is the positivity of the exploitation rate. It can be shown that the positivity of the exploitation rate, together with the assumption that all industries in the economy have the same value-composition of capital, i.e. that the ratio of constant to variable capital is the same in every production process, implies that the equilibrium profit rate is positive and —what is much more— that equilibrium prices turn out to be proportional to values. Nevertheless, it is possible to have in a simple economy both a positive uniform profit rate and a positive uniform rate of profit without all industries having the same value-composition of capital. In such a case, the equilibrium price systems deviate from the value systems, in the sense that they are found to be nonproportional to the latter.

We have not introduced yet the important concept of explotation rate. In order to introduce it, it is important to define the surplus-value of a production process. This is the difference between the amount of live labor expended in the process and the value of the goods the workers can afford with their salary. The value represented by the salary paid in production process i ($i = 1, \ldots, l$) (operated at unitary level), the so-called necessary labor, is just $v_i = (\lambda_{k+1}b_{k+1} + \cdots \lambda_l b_l)x_i$. On the other hand, the total amount of live labor expended in this process is x_i. Hence, the surplus-value of the process is $x_i - v_i$ and so the rate of exploitation, as defined by Marx in C,[18] is the ratio of surplus-value to necessary labor:

$$\varepsilon_i = \frac{x_i - v_i}{v_i}.$$

Marx referred to number ε_i as "an exact expression for the degree of exploitation of labor-power by capital, or of the worker by the capitalist".[18] For this reason, the rate of surplus-value is also called 'the rate of exploitation'. In general, the rate of exploitation may vary from one production process to another, but in the special case of a simple economy this rate turns out to be uniform for all production processes. This is the content of the next theorem.

THEOREM 4: *In a simple economy the exploitation rate is uniform.*

Proof: Consider any two production processes $\tilde{\mathbf{x}}_i$ and $\tilde{\mathbf{x}}_j$ $(i, j = 1, \ldots m)$ and let α be the number x_j/x_i. It is obvious that $\alpha > 0$ and that $v_j = \alpha v_i$. Therefore,

$$\begin{aligned}
\varepsilon_j &= \frac{x_j - v_j}{v_j} \\
&= \frac{\alpha(x_i - v_i)}{\alpha v_i} \\
&= \varepsilon_i. \quad \square
\end{aligned}$$

Marx defined the constant capital of a labor process as the value of the means of production,[19] i.e. the constant capital of production process $\tilde{\mathbf{x}}_i$ $(i = 1, \ldots, m)$ is the number c_i as determined by

$$c_i = \underline{x}_{1i}\lambda_1 + \cdots + \underline{x}_{ki}\lambda_k.$$

On the other hand, the surplus-value is the number s_i given by

$$s_i = x_i - v_i,$$

which represents the amount of new value added to the capital invested in the process. Clearly, $s_i = \varepsilon v_i$, and so the following proposition is seen to be true.

THEOREM 5: *The value λ_i of the product of process $\tilde{\mathbf{x}}_i$ $(i = 1, \ldots, m)$ can be expressed as*

$$\lambda_i = c_i + v_i + \varepsilon v_i.$$

By virtue of Theorem 5 we can write the value equations for capital and wage goods as

(10) $\Lambda = \Lambda\mathbf{A} + \Lambda\mathbf{BL} + \varepsilon\Lambda\mathbf{BL}$

where

$$\Lambda = [\lambda_1 \cdots \lambda_l]$$

is the vector of values of capital and wage goods industries. Analogously, the value equations for luxury goods can be written as

(11) $\Lambda' = \Lambda\mathbf{A}' + \Lambda\mathbf{BL}' + \varepsilon\Lambda\mathbf{BL}'$

where

$$\Lambda' = [\lambda_{l+1} \cdots \lambda_m]$$

is the vector of values of luxury goods,

$$\mathbf{A}' = \begin{bmatrix} \underline{x}_{1l+1} & \cdots & \underline{x}_{1m} \\ \vdots & & \vdots \\ \underline{x}_{kl+1} & \cdots & \underline{x}_{km} \end{bmatrix}$$

is the matrix of inputs of luxury goods industries, and

$$\mathbf{L}' = [x_{l+1} \cdots x_m]$$

is the vector of labor inputs of the same industries.

Now we are in position to establish a condition that is both necessary and sufficient to guarantee profitable wages and prices for all the industries in a simple economy. As I said above, this is precisely the FMT.

THEOREM 6: (The Fundamental Marxian Theorem for a simple economy). *In a simple economy, there exists a price system at which the rate of profit of every process is positive iff the rate of exploitation ε is positive.*

Proof: Assume first that the rate of profit is positive for every production process, i.e. assume that equations (8) and (9) hold. Since **p** is positive,

$$\mathbf{pCy} < \mathbf{py}$$

for any positive column vector **y**. This would not be possible if **C** were not productive, for the following reasons. Since **C** is semireproducible, there is a positive vector **z** such that $\mathbf{Cz} \leqq \mathbf{z}$, i.e. such that either $\mathbf{Cz} = \mathbf{z}$ or $\mathbf{Cz} \leq \mathbf{z}$. But the first case is impossible because it implies

$$\mathbf{pCz} = \mathbf{pz},$$

contradicting what was established above. Hence $\mathbf{Cz} \leq \mathbf{z}$. Thus there is a positive vector **y** such that (see equation 10)

$$\Lambda\mathbf{Cy} + \varepsilon\Lambda\mathbf{BLy} = \Lambda\mathbf{y} > \Lambda\mathbf{Cy}$$

and so $\varepsilon\Lambda\mathbf{BLy} > 0$. It follows that $\varepsilon > 0$.

Assume now that $\varepsilon > 0$ and let $\mathbf{p} = \Lambda$, $\mathbf{p}' = \Lambda'$. Since $\mathbf{pB} = \mathbf{w}$, $\Lambda\mathbf{L} = w\mathbf{L} > \mathbf{0}$ and $\Lambda\mathbf{L}' = w\mathbf{L}' > \mathbf{0}$. Therefore, equations (10) and (11) yield

$$\mathbf{p} = \mathbf{pC} + \varepsilon w\mathbf{L} > \mathbf{pC}$$

and

$$\mathbf{p}' = \mathbf{pC}' + \varepsilon w\mathbf{L}' > \mathbf{pC}'.$$

This establishes that the profit rate of each process is positive. □

The FMT establishes that a positive rate of explotation is both necessary and sufficient for the existence of a price system at which all production processes in the economy are profitable. It does not establish, however, that this price system has to be exactly the one corresponding to the unique equilibrium rate of profit whose existence was established in Theorem 3. In other words, the profit rate of one production process is not necessarily equal to that of another.

The *transformation problem* is the problem of establishing a law relating labor-values and prices in such a way that the latter can be derived out of the former. The transformation problem is the problem of obtaining the equilibrium prices out of the value systems by means of a regular nomological pattern. In the special case in which the constant composition of capital is the same in all the economy, the transformation problem can be solved in a straightforward way: as a matter of fact, it can be shown that equilibrium prices are proportional to values iff all industries in the economy have the same value-composition of capital, i.e. if the ratio of constant to variable capital is the same in every production process. This is the gist of this, the last theorem of the chapter.

THEOREM 7: *In a simple economy, equilibrium prices are proportional to values iff all industries in the economy have the same value-composition of capital.*

Proof: Assume first that equilibrium prices are proportional to values. That is to say, there is a positive α such that $p_i = \alpha\lambda_i$ for $i = 1, \ldots, m$. Let us write C_i for the sum in terms of equilibrium prices: $p_1\underline{x}_{1i} + \cdots p_k\underline{x}_{ki}$ and V_i for the amount wx_i, also determined in terms of equilibrium prices. Then we have

$$
\begin{aligned}
r &= \frac{\mathbf{p}\widehat{\mathbf{x}}_i - wx_i}{wx_i + \mathbf{p}\underline{\mathbf{x}}_i} \\
&= \frac{p_i - (C_i + V_i)}{C_i + V_i} \\
&= \frac{\lambda_i - (c_i + v_i)}{c_i + v_i} \\
&= \varepsilon \frac{v_i}{c_i + v_i}.
\end{aligned}
$$

Therefore, c_i/v_i is constant and equal to $r^{-1}\varepsilon - 1$.

Assume now that c_i/v_i is constant for $i = 1, \ldots, m$, and set

$$
r = \varepsilon \frac{c_i}{v_i}.
$$

Then we have

$$(1 + r)(c_i + v_i) = c_i + v_i + r(c_i + v_i)$$
$$= \lambda_i - \varepsilon v_i + r(c_i + v_i) \qquad \text{(Theorem 5)}$$
$$= \lambda_i - \varepsilon v_i + \varepsilon v_i$$
$$= \lambda_i$$

Hence, we have just obtained the equations

$$\Lambda = (1 + r)\Lambda\mathbf{C} \qquad \text{and} \qquad \Lambda' = (1 + r)\Lambda\mathbf{pC}'.$$

Theorem 3 implies then that Λ is proportional to $\mathbf{p}$ and Λ' is proportional (by the same factor) to $\mathbf{p}'$. $\square$

Morishima has shown that important results established by Marx in the third volume of C hold with certain modifications and additional restrictive assumptions, and so Morishima thinks that Marx was "succesful" in the transformation problem. Morishima thinks that Marx motivation to tackle this problem was to show that "individual exploitation and individual profit are disproportional unless some restrictive conditions are imposed".[20] According to Morishima,

> it is clear that the transformation problem has the aim of showing how 'the aggregate exploitation of labor on the part of the total social capital' is, in a capitalist economy, obscured by the distortion of prices from values; the other aim is to show how living labor can be the sole source of profit [...] Marx [...] was very succesful in the transformation problem.[21]

Nevertheless, leaving aside the fact that such additional assumptions are rather unrealistic, the gist of the labor theory of value was in the first place to show that the Law of Value holds true. This law asserts that the magnitude of value of commodities *regulates* (in a rather interesting, nontrivial sense) the proportions in which they exchange. Now, it is true that many interpreters of Marx have taken this "regulating" as demanding the proportionality of equilibrium prices and values. Böhm-Bawerk, one of the most outstanding critics of Marx in the nineteenth century, claimed that the Law of Value asserts "and for all that precedes cannot assert, but that commodities exchange among themselves in proportion to the average socially necessary labor-time incorporated to them".[22] Thus far we have seen that if this proportionality is understood as the proportionality of values and equilibrium prices, within the conceptual framework of the prototype of MTV, then the Law of Value is seen to fail, except in the uninteresting case of equal value-composition of capital, according to Theorem 7. We shall see if it is possible to provide a formulation of the Law of Value that circumvents this fundamental difficulty. In this connection, the problem is to make sense of the idea that value "regulates" prices. This is one of the foundational problems of MTV. We shall formulate it and all the others in the following chapter.

These results are sufficient for our purposes of examining the foundational problems of MTV. It can be seen that MTV is a serious important theory that deserves attention by scientists and philosophers of science. Now I shall proceed to the next chapter, where the foundational problems of the theory will be presented and analyzed in detail. We shall see that these problems arise mainly out of the effort to provide a general quantitative concept of value that makes possible a correct general formulation of the Law of Value, as well as from the attempt to generalize the assumptions that enabled us to prove the interesting theorems of MTV for the prototype.

Chapter 2

THE PROBLEM OF FOUNDATIONS

What is the content of the problem of foundations and its importance? We have seen that even the prototype of MTV —which is a quite idealized version of a capitalist economy— is interesting as a particular nonexistent artificial system in which the structure of the capitalist economy can be studied. I will discuss the scientific relevance of such systems in the chapter on the dialectical method. In the meantime, we can say that the simple economy studied in the previous chapter is useful because it sets the stage for our raising of certain questions which must be of interest to anybody interested in the real workings of modern market economies. The two most important of these are, clearly, whether the idealizing assumptions that define simple systems can be generalized in such a way that general laws can be formulated that obtain in those economies. In particular, is it possible to prove that a quantitative concept of value can be defined in more general structures? And what about the Law of Value? Is it possible to provide a more precise formulation of it, a formulation that can be shown to hold not only in the case of simple commodity production with equal composition of constant capital, but also in modern full-blown capitalist economies? These are the main foundational problems of MTV. The present chapter intends to give a detailed and rigorous formulation of them.

2.1   THE SENSE AND IMPORT OF THE LAW OF VALUE

The aim and goal of a theory of value is to give a satisfactory explanation of the fact that, in the market, commodities gravitate around certain prices. *Why these commodities gravitate around these prices and cannot go much below or much above them?* This is one way of asking the fundamental problem of a theory of value. Roughly speaking, the value a commodity has is "what it costs to get it". Common sense has a good appraisal of the value of commodities in daily life, as it is evinced by the fact that nobody expects to obtain in the market a Mercedes Benz for ten American dollars. This strong sense concerning the value of things suggests that there must be something objective, some reason or reasons why things happen to be as costly as they are. The labor theory of value intends to explain the cost of things in terms of social labor. *Essentially*, a labor theory of value must explain the movement of prices in terms of social labor, and that is why the Law of Value —which asserts that, in the very least, labor "regulates" the cost or value of commodities— constitutes in itself the very core of the labor theory of value. As the Spanish philosopher Francisco Álvarez (1986) has pointed out, the Law of Value is actually the law that defines MTV, very much as Newton's Second Law defines his mechanics. Hence, in a sense, Marx's Law of Value *is* his labor theory of value. This is the reason why we, as philosophers of science, must take seriously the claims made by Marx concerning the regulative role of value in price formation, and try to solve the conceptual difficulties that it poses. Accordingly, I shall take here at its face value Marx's formulation of the Law of Value, in the sense that the magnitude of value of commodities *regulates* the proportions in which they exchange. I shall discuss Böhm-Bawerk's criticism of this law in order to conclude that this criticism is right but not quite right. The blame for the blast Marx gets from this critic lies no doubt on Marx own hesitations, obscurities, inconsistencies and final methodological decision (all of which we followed carefully in the first chapter) that led him to claim that values are to be determined independently of the market. This turned out to be a blind alley that

got Marx —despite Morishima's sympathetic interpretation—
into the contradictions crudely pointed out by Böhm-Bawerk.
We shall see that these contradictions are the result of the basic
opposition introduced by Marx between the sphere of commodity production and the sphere of distribution, an opposition implicitly contained in the opposition between value as determined
solely in the sphere of production and value as determined solely
in the sphere of distribution (in the market). We shall see that
these are two arrested moments of a dialectical unity, which is
the concept of *abstract labor* taken in a sense rather different to
the one Marx gave to it in C.

Even within a simple economic system (in the sense of chapter
1), there seems to be a contradiction between the account Marx
gives of value —especially of the Law of Value— in the first volume of C and his acknowledging that if commodities were sold
by their values then every firm or industry would tend to have
a different profit rate, which contradicts his own remark that in
a real capitalist economy the capitals move from one sphere of
production to another, looking for the highest return rate, which
induces a generalized tendency within the economy toward the
equalization of the profit rate. Clearly, if the Law of Value is
taken to mean that commodities are (tendentially) sold at their
values, then this contradiction is nothing but a good Popperian
refutation of the Law of Value. The way out of this dilemma for
Marx and his sympathetic interpreters has been twofold: (1) On
one hand, when Marx saw that his early formulation of the Law
of Value contradicted the results he had arrived at in volume 3
of C, he began to water-down the import of the Law of Value,
reducing it to the claim that value "regulates the exchange relations" in a rather imprecise sense. (2) On the second, Marx and
his followers —notably Engels in the Preface and Supplement
to volume 3 of C— have interpreted the distressing results obtained by Marx in volume 3 as showing that the Law of Value
as formulated in volume 1 is valid only in a society of "simple
commodity production":

The exchange of commodities at their values, or at approximately these values, thus corresponds to a much lower stage of development than the exchange at prices of production, for which a definite degree of capitalist development is needed.[1]

(The reader must notice that what Engels calls 'simple commodity production' is *not* what *I* called 'simple Marxian economy'. Our simple economy is fully capitalist, even if rather idealized and simplified in this sense. Engels' simple commodity production economy is supposed to be a stage historically previous to developed capitalism, but in fact we have seen (see Theorem 7 of chapter 1) that the Law of Value holds exactly in our capitalist simple economy, under the assumption of an equal organic composition of capital.)

Before considering Böhm-Bawerk's criticisms, we shall see the different characterizations of the Law of Value provided by Marx along the third volume of *C*. Noticing that the equalization of profit rates contradicted his original formulation of that law, Marx suggested that even if prices are not proportional to values, yet

the sum of prices of production for the commodities produced in society as a whole —taking the totality of all branches of production— is equal to the sum of their values.[2]

This is Marx's first attempt to cope with the problem. A couple of pages later, complaining that the number of hours that the worker must work in order to produce his means of subsistence "is distorted by the fact that the production prices of the necessary means of subsistence diverge from their values", Marx had asserted that this "is always reducible to the situation" that

whenever too much surplus-value goes into one commodity, too little goes into another, and that the divergences from value that obtain in the production prices of commodities therefore cancel each other out.[3]

And after this he immediately adds as a conclusion the following statement:

With the whole of capitalist production, it is always only in a very intricate and approximate way, as an average of perpetual fluctuations which can never be firmly fixed, that the general law prevails as the dominant tendency.[4]

Unfortunately, if this conclusion was intended by Marx to summarize his first attempt to deal with the transformation problem, it is not well established, because —as Morishima has shown— the sum of prices is equal to the sum of values only in the uninteresting case in which both the rate of exploitation and the rate of profit are zero.[5]

In the following chapter, Marx seems to try yet another way of formulating the Law of Value:

Whatever be the ways in which the prices of different commodities are first established or fixed in relation to one another, the Law of Value governs their movement. When the labor-time required for their production falls, prices fall; and where it rises, prices rise, as long as other circumstances remain equal. [...] In whatever way prices are determined, the following is the result: [...] The Law of Value governs their movement in so far as reduction or increase in the labor-time needed for their production makes the price of production rise or fall. [...] The average profit, which determines the prices of production, must always be approximately equal to the amount of surplus-value that accrues to a given social capital as an aliquot part of the total social capital. [...] Since it is the total value of the commodities that governs the total surplus-value, while this in turn governs the level of average profit and hence the general rate of profit —as a general law or as governing the fluctuations— *it follows that the Law of Value regulates the prices of production.*[6]

Unfortunately, the part of the former proposition which asserts that prices move as the labor-time required for their production moves is trivially true: If the labor-time required for the production of commodities augments (diminishes), the salaries also are increased (decreased) and this in turn implies a price increment (decrement) (use formula (5) of chapter 1). We shall deal below with the clause that follows thereafter.

In the same chapter Marx proposes still another way of formulating the Law of Value when he says that

the relationship between demand and supply does not explain market value, but it is the latter, rather, that explains fluctuations in demand and supply.[7]

Here the formulation is sufficiently general as to elude criticism. But in chapter 18 Marx specifies the claim, denying the
market any role in the determination of equilibrium prices.
There he sees the apparent determination of price by the turn-
over of commercial capital as a distorting effect that obliterates
the inner connection of the rate of profit with the formation of
surplus-value, creating the "illusion" that

> the circulation process as such determines the prices of commodi-
> ties, and that this is within certain limits independent of the process
> of production.[8]

Clearly, this is very suggestive of the role Marx attributes to
the Law of Value even at this very advanced stage of C, since
the former complaint clearly implies that any determination of
prices independently of the sphere of production, even within
certain limits, is "mere illusion", that is, (equilibrium) prices must
be determined by value. And this is indeed what Marx had said
a few lines above, where he claimed that

> while a closer consideration of the influence of turnover time on
> value formation in the case of the individual capital leads back to
> *the general law and the basis of political economy*, viz. that commodity
> values are determined by the labor-time they contain, the influence
> of the turnover of commercial capital on commercial prices exhibits
> phenomena which, in the absence of a very far-reaching analysis of
> the intermediate stages of the process, seem to presuppose a purely
> arbitrary determination of prices.[9]

Hence, the Law of Value appears at this point —despite all the
difficulties pointed out by Marx in previous chapters— as "the
general law and basis of political economy", as a law asserting
that values or prices are determined by labor-time.

On chapter 37, Marx provides yet another characterization of
the Law of Value, this time as a regularity obtaining between values and production prices whenever the social division of labor is

proportionate to the social needs, i.e. whenever society allocates the required amounts of social labor to the different branches of production, so that the different social needs are satisfied:

> If [the social division of labor] is in due proportion, products of various types will be sold at their values (at a further stage of development, at their prices of production), or at least at prices which are modifications of these values or production prices *as determined by general laws. This is in fact the Law of Value as it makes itself felt, not in relation to the individual commodities or articles, but rather to the total products at a given time of particular spheres of social production autonomized by the division of labor*; so that not only is no more labor-time devoted to each individual commodity than necessary, but out of the total social labor-time only the proportionate quantity needed is devoted to the various types of commodity.[10]

Morishima followed the suggestion contained in the italicized fragment of this quotation, finding a more general law of transformation of values into prices proportional to the equilibrium prices. Under these prices, the sum of prices of production for the commodities produced in society as a whole is equal to the sum of their values and, moreover, the total surplus-value equals the total profits. Unfortunately, this holds under the rather restrictive assumption that industries are linearly dependent (i.e. that the matrix

$$\begin{bmatrix} A_I, & A_{II} \\ BL_I, & BL_I \end{bmatrix}$$

is singular) and so it hardly constitutes the desired general formulation of the Law of Value.

In chapter 49 Marx confronts the additional "confusion" derived from the transformation of surplus-value into profit and rent, but he insists that the relations the different agents of production have to these particular components "in no way alter the value determination and its law":

> Just as little is the Law of Value affected by the fact that the equalization of profit, i.e. the distribution of the total surplus-value among the various capitals and the obstacles that landed property places

in the way of this (in absolute rent), gives rise to governing average prices for commodities that diverge from their individual values. This again affects only the addition of surplus-value to the various commodity prices; it does not abolish surplus-value itself, nor the total value of commodities as the source of these various price-components.[11]

Finally, in chapter 51 of volume 3, the next to last of C, Marx thinks of the Law of Value as asserting the regulation of "the total production by value":

the [...] characters of the product as commodity and the commodity as capitalistically produced commodity give rise to the entire determination of value and the regulation of the total production by value.[12]

Hence, we can see that throughout volume 3 Marx provides several characterizations of the Law of Value, being the constant that value determines or regulates the exchange relations. Despite the serious difficulties encountered by Marx in his quest for a more definite formulation of the Law of Value, he nevertheless was convinced, up to the last chapter of C, that value somehow regulated the production prices of commodities. Despite the insights provided by the former remarks on the Law of Value, the problem of specifying the way in which value regulates prices is still open. It seems to me that Morishima's sympathetic claim that Marx was "very succesful" in the transformation problem is excessively optimistic. For even Morishima's revised formulation of Marx's solution to that problem depends on unduly and unnaturally restrictive assumptions. Rather, it seems that Marx thought, while he was writing the first volume of C, that he had unveiled the secret of the value of capitalistically produced commodities only to face sometime later —while he was writing volume 3— the distressing appearence of serious difficulties to sustain the Law of Value in the way he seems to have thought it held when he wrote volume 1. I challenge the Marxian scholars to prove that Marx was aware, when he formulated his Law of Value in volume 1, that it was valid only for a supposed "simple commodity production" economy. Quite on the contrary, I am inclined to think that there is a real contradiction between Marx's formulation of the Law of Value in volume 1, and the hard facts he discovered later while writing volume 3. Nowhere in *C* was Marx able to provide a solution to this contradiction; we shall see that his remarks on the Law of Value along the third volume of *C* do not really solve the problem. As Böhm-Bawerk stressed, if in reality we observe that there is a tendency toward equalization of profit rates, and that the composition of capitals is quite diverse, then Marx's formulation of the Law of Value appears to be incorrect.

Böhm-Bawerk analyzed the main arguments advanced by Marx along volume 3 in order to solve the former difficulty. These are four (see the quotations above):

(1) The sum of prices of production for the commodities is equal to the sum of their values.

(2) When the labor-time required for the production of commodities falls, prices fall; and where it rises, prices rise, as long as other circumstances remain equal.

(3) The Law of Value as given in volume 1 of *C* is exactly true in the case of "simple commodity production" economies.

(4) In an advanced capitalist economy, the Law of Value regulates at least indirectly and ultimately the production prices, since it is the total value of the commodities that governs the total surplus-value, while this in turn governs the level of average profit and hence the general rate of profit.

In connection with the first argument, we have said that Morishima has established that the sum of prices of production for the commodities is equal to the sum of their values only under the arbitrary assumption that industries are "linearly dependent". Hence, *contra* Böhm-Bawerk, this proposition is far from being a mere tautology. I think that nevertheless this author raises two serious criticisms to it. I think that Böhm-Bawerk's contention that Marx confuses a mean among fluctuations and a mean between constant and fundamentally unequal magnitudes is right. Marx's *fiesta* of averages does not solve at all the problem, because one thing is to establish that prices fluctuate around values determined as in chapter 1, and quite another to say that they fluctuate around a certain mean. As Böhm-Bawerk puts it, "that mean has a completely different meaning or, more exactly, it is entirely meaningless for our law".[13] No matter how tough may this criticism sound, the main criticism is yet more subtle and corrosive. After asking "What is the mission of the "Law of Value"?", Böhm-Bawerk replies —quite correctly: "Evidently, only that of clarifying the exchange relation of goods as it is observed in reality".[14] This is clearly seen in the original formulation of the law, according to which equilibrium prices are proportional to values (and we observe in reality a tendency toward equilibrium prices, i.e. to prices determined according to a uniform rate of profit). Hence, Böhm-Bawerk is right in pointing out that even if the claim that globally prices are equal to values were true, it would be fully irrelevant, since "at any rate, certainly you do not answer in political economy the question of which is the exchange relation of commodities by indicating the sum of prices they get altogether".[15] It seems that in this point Böhm-Bawerk is right, and Marx's answer appears as a desperate effort to tackle with the problems the Law of Value had begun to confront, since his answer is not even relevant to the point.

As I said above, proposition (2) is trivially true. If we keep all circumstances equal, then it is obvious that an increase (decrease) of labor-time implies and increase (decrease) in the production price of the commodity. For in such a case it is necessary to invest more in salaries. Clearly, Böhm-Bawerk is quite right when he claims that this argument is therefore useless to show that what *exclusively* determines the magnitude of the value of any article is the amount of labor socially necessary, or the labor-time socially necessary for its production. For the argument shows only that labor-time is but *one* determinant cause of prices:

> Evidently, one could affirm that *this* law dominates the movements of prices only if a (permanent) change of prices could not be op-

erated or conditioned by any other cause but the variation of the magnitude of labor-time.[16]

Böhm-Bawerk attacks argument (3) by showing that in a "simple commodity production" economy workers (who own the production means) would require different production times, and so each must wait for different periods of time to receive compensation. Since it is a defining assumption of our simple Marxian economy that all products have the same period of production, that assumption could be extended also to characterize in its ideal purity a simple commodity production system. Hence, Böhm-Bawerk's criticism can be raised also against the prototype of MTV. We shall deal with this type of criticisms, directed against idealizing assumptions, in the following section. At this point the interesting question is whether it can be historically proven the existence of such systems, whether it is possible to find traces of societies of that kind. Böhm-Bawerk's conclusion is that "in reality, it is not possible to discover this traces in any place whatsoever, neither in the historical past nor in the present".[17] According to Böhm-Bawerk, the Law of Value has never exerted, and could not have exerted, a real supremacy even in primitive conditions, due to the fact that the phenomenon of concurrency appeared already at the earliest stages of capitalist production (contrary to what Engels claims, the simple commodity production economies are capitalist).

Böhm-Bawerk considers that his rebuttal of arguments (1)-(3) establishes that three formulations of the Law of Value, intended to show that that law had validity under certain restrictions, are mistaken and miss the target: The Law of Value fails to hold even under such conditions. The fourth argument is of a different kind, since it does not intend to show that the law holds under special circumstances, but rather to establish that the law is one of unrestricted validity, only that its form is not like the one Marx had enunciated in the first volume of C. Böhm-Bawerk finds the most precise formulation of the generalized Law of Value in an already quoted passage which reads:

The average profit, which determines the prices of production, must always be approximately equal to the amount of surplus-value that accrues to a given social capital as an aliquot part of the total social capital. [...] Since it is the total value of the commodities that governs the total surplus-value, while this in turn governs the level of average profit and hence the general rate of profit —as a general law or as governing the fluctuations— *it follows that the Law of Value regulates the prices of production.*

Böhm-Bawerk analyzes the links of the former chain of reasonings, showing that in each one of these links —total value - total surplus-value, total surplus-value - average profit, average profit - general rate of profit, general rate of profit - prices of production— Marx fails to see that one factor additional to labor-time is a concomitant determinant of the following element in the sequence. Hence, Böhm-Bawerk concludes that there is a real contradiction between *all* the formulations of the Law of Value provided by Marx and the facts:

> the Law of Value pretends that only the amount of labor determines the exchange relations; the facts prove, on the contrary, that it is not only the amount of labor, or factors homogenous to the same, which determine the exchange relations.[18]

It seems to me that Böhm-Bawerk's criticism of the formulations of the Law of Value given by Marx is solid and cogent. It is beyond doubt that, at the very best, Marx only was able to show that labor-value is but one of the multiple concomitant factors determining the equilibrium or "production" prices. I shall take these criticisms seriously, as constituting a harsh attack on the very foundations of MTV. We shall see that Böhm-Bawerk was right in claiming that the formulations of the Law of Value provided by Marx conflict with the facts, but not quite right in the sense of not being able to notice that perhaps *other* formulations are in agreement with them.

In the second section we shall complicate the panorama even more. The former criticisms of the Law of Value hold even within the context of a simple Marxian economy. We shall ask whether it is possible to drop at least some of the assumptions, or even better, if it is possible to obtain a completely general formulation of MTV and, within this new formulation, whether it is possible to provide a new formulation of the Law of Value that overrides the correct criticisms advanced by Böhm-Bawerk.

2.2 THE PROBLEM OF GENERALIZING THE PROTOTYPE

In spite of the essentially failed attempts to provide a cogent formulation of the Law of Value within our simple Marxian economy, we were nevertheless able to obtain a number of substantial results. The main of these is perhaps the Fundamental Marxian Theorem for such a kind of economies. Indeed, these results were obtained under the assumption that certain conditions —some of which are quite restrictive— hold. Our task in the present section will be to identify all of these conditions and to explore the question whether it is possible to obtain the same results —or even better results, since we also want an acceptable formulation of the Law of Value— by canceling those assumptions which are restrictive. We shall see that the problem of providing an acceptable formulation of the Law of Value is deeply connected with the problem of finding an adequate set of nonrestrictive assumptions that imply the desired results.

The prototype of MTV is built upon assumptions which are general and others that are restrictive. The restrictive assumptions are the following:

(S1) *All capital goods have the same period of rotation*

(S2) *There are no heterogeneous concrete labors, i.e. all labor-power is homogeneous*

(S3) *There is no choice of techniques*

(S4) *There is no joint production*

(S5) *The technology yields constant returns to scale, and so goods and production processes are infinitely divisible*

(S6) *There is a fixed consumption bundle for all workers, and so the demand structure of the proletariat is very rigid*

(S7) *Workers do not save and so their salary is exactly suficient to acquire the consumption bundle*

(S8) *The economy is closed*

Having clearly before us the former assumptions, the problem of generalizing the protoype can be precisely formulated now as the problem of providing an adequate characterization of a capitalist market economy, a characterization powerful enough to entitle us to prove the existence of numerical labor-values, to prove the important theorems of the MTV, and above all to provide a plausible formulation of the Law of Value. We shall present now an historical overview of the efforts that have been made to get rid of the former assumptions.

2.3 A CONCISE HISTORY OF MTV

The first step toward a mathematical formulation of MTV was given by Wassily W. Leontief in Part II of *The Structure of American Economy 1919-1929*, published for the first time in 1941. In that book Leontief introduced a "theoretical scheme" based on what later came to be known as "Leontief matrices", precisely because of their appearing in this work of Leontief's. For the sake of historical accuracy, however, it is fair to say that the first one in introducing such matrices to theorize in the field of economics was Vladimir Karpovich Dmitriev (1868-1913) in his essay "The Theory of Value of David Ricardo", published originally in 1898. At the beginning of this essay Dmitriev considered the following question: "How it is possible to calculate the amount of labour expended for the production of a given economic good from the very beginning of history, when man managed without capital, down to present time?"[19] After pointing out that the amount of labor expended on the production of a given product may be determined "without such historical digressions",[20] Dmitriev produced a full-fledged input-output system which is identical to Leontief's labor input-output system. According to Nuti (1974), the analytical apparatus provided by Leontief four decades later adds two things: (i) a method for the actual computation of the solution, namely the inversion of the matrix $(I - A^T)$, where I is the identity matrix and A^T is the transpose of the matrix of technical coefficients; and (ii) the generalization of the notion of full input (i.e. direct and indirect input requirements) from labor to other production inputs.[21]

It is very difficult to know whether Leontief actually knew Dmitriev's work before the publication of his famous book. Be that as it may, the objective identity of Leontief's labor input-output system with that of Dmitriev's proves that Leontief was not the first one to devise it, although it also points out to something which is more important, to wit, the Ricardian lineage of Leontief's theoretical apparatus.

The starting point of the mathematical formulation of the prototype of MTV is Leontief's theoretical apparatus as presented in his book of 1941, but —as we saw in chapter 1— this prototype is already sketched in the first two volumes of C. One of the first persons ever in making explicit the assumptions of Leontief's theoretical apparatus was Georgescu-Roegen (1950), but perhaps the first writer to draw attention to the fact that such apparatus is grounded upon a labor theory of value was Burgess Cameron (1952). Cameron showed that the proposition 'the [Leontief] price of a commodity (in terms of the wage rate) will equal the number of man-hours to produce it' is derivable within Leontief's system.[22] Morishima and Seton (1961) took for granted that in obtaining the former result Cameron also established

that in a Leontief model the price of a commodity in terms of labor ("wage price") will equal its Marxian "value" under certain conditions which include (i) competitive long-run equilibrium, i.e., zero profits in each sector, and (ii) perfect divisibility of the economy into "primitive sectors", i.e., sectors producing single homogeneous commodities.[23]

Now, since assumption (i) in destroying the "surplus" (i.e. zero profit) clearly "reduces the discovered equality to a merely formal one", Morishima and Seton's purpose was "to inquire into the general relationship between Leontief price and Marxian value when both assumptions (i) and (ii) are relaxed".[24] In doing so, Morishima and Seton developed a less idealized version of the theory and —within this still very idealized version— they were able to obtain mathematically for the first time (their own versions of) two classical Marxian results, namely, (1) "The rate of exploitation will always exceed the rate of profit"; and (2) "A Marxian type of 'economic progress' (where capital accumulation steadily reduces the share of wages in the total costs of all sectors) will normally entail a fall in the rate of profit, unless accompanied by a countervailing in the rate of exploitation".[25] Result (1) is close to what we have called the Fundamental Marxian Theorem. On chapter 1 we formulated this theorem in the following terms (see Theorem 6): There exists a price system at which the rate of profit is positive if and only if the rate of exploitation is positive. Okishio (1963) provided an independent proof of a version of the theorem, which is even closer to this formulation, using assumptions similar to those of Morishima and Seton.[26] The assumptions and structure of the classical model appear in a rather explicit form in this paper of Okishio's, but the first thorough examination and description of the same model appeared one decade later in what is now one of the classics of the Marxian literature, Morishima's *Marx's Economics*.[27] In this book there is an exhaustive list of the assumptions of the classical model and detailed proofs of all the central theorems. A result which is particularly important from the point of view of the philosophy of science is the proof of the existence of unique positive solutions of the Leontief labor-value equations and a proof of the equivalence between value understood as socially necessary labor-time and value understood as the sum of the amount of direct labor and "congealed" labor embodied in the means of production which is transferred to the product.[28] Using these precisely defined Marxian values, Morishima developed a considerable amount of Marx's economic theory in a scientific rigorous way, although of course within the limitations imposed by the idealized assumptions on which the proofs concerning the existence of unique positive values are based. These assumptions (enlisted in the previous section) were criticized by Morishima in the same book, whose conclusion is that MTV has to be abandoned and replaced by a new theory that combines features of MTV and of Von Neumann's economic theory.

Morishima (1974) addressed the problem of constructing a new Marx-Von Neumann theory of value. Unlike the prototype, this new theory drops assumption (S1), that all capital goods have the same period of rotation, and makes possible a better treatment of capital age-structure problems when the time factor is introduced (which are intractable within the prototype). To achieve this, it admits both joint production and choice of techniques, thus droping also assumptions (S3) and (S4). Morishima was able to prove the Fundamental Marxian Theorem within this Marx-Von Neumann theory, using "optimum" values instead of "actual" values. If actual values are obtained by calculating the embodied labor contents of commodities on the basis of the prevailing production coefficients, the optimum values are shadow prices determined by a linear programming problem which is dual to another linear problem for the efficient utilization of labor. Although optimum values are not necessarily unique, the rate of exploitation is well defined and —as I said before— the Fundamental Marxian Theorem can be proved, provided that labor is assumed to be homogeneous.

On the same line of Morishima's, John E. Roemer (1980, 1981) produced a series of more general theories and derived the existence of Marxian equilibria from the assumption that the firm's behavior consists of maximizing profits given a set of possible production processes and certain restrictions of capital availability. In Roemer's theories the value of a bundle of commodities is defined as the minimum labor required to produce the bundle, given the technological possibilities of the economy; thus, Roemer's definition is analogous to that of Morishima's, the difference being that according to Roemer values are not necessarily determined by a *linear* program. In Roemer's theories the exploitation rate is well defined for each production process and a more general version of the Fundamental Marxian Theorem is proved. In Roemer's models, like in the one built by Morishima on Von Neumann's, the role of the exploitation of workers is clarified as a condition for the growth and reproduction of the economy. Roemer's models are fairly general but they are still based on assumptions (S2), (S5)-(S8). Moreover —as it is also the case for the Marx-Von Neumann theory— no satisfactory account of the Law of Value can be given.

In particular, assumption (S2), the assumption that labor is homogeneous, is very restrictive. According to Elster (1985),

> the presence of genuinely and irreducibly heterogeneous labor is a major stumbling block for Marxist economics. If taken seriously, it prevents the labor theory of value from even getting off the ground, since the basic concepts cannot be defined.[29]

It seems to me that this remark of Elster's must be taken very seriously, because one of the most outstanding traits of capitalism is precisely a very rich division of labor. We shall see that with this remark Elster has really hit the core of the foundational problems of MTV. It will be apparent that in order to deal with this problem it will be necessary to return to the very basis of MTV as given by Marx in the first chapter of C.

Before attempting to address the problem of heterogeneous labor, I would like to close the present chapter by presenting in some mathematical detail the achievements of Morishima and Roemer.

Leaving aside the intrinsic limitations of (S3) and (S4), Morishima found that it is impossible to provide a consistent treatment of aging capital goods within these assumptions. Even assuming that (S3) and (S4) are satisfied by processes producing brand new goods, serious difficulties arise in the treatment of the wear and tear of fixed capital goods. Morishima did show that

> There will not be universal consistency between 'the replacement of the wear and tear portion of the value in the form of money' and 'the replacement of fixed capital in kind', unless we can get rid of the neoclassical method of depreciation and obey the Von Neumann golden rule in the evaluation of capital costs.[30]

According to Von Neumann's golden rule, used capital goods must be treated as by-products in the labor process. But this is tantamount to the rejection of (S3) and (S4), and so the adoption of Von Neumann's treatment of capital goods implies the subversion of an important portion of the very foundations on which the existence of labor-values had been made to depend in the construction of the prototype.

In his 1974 paper Morishima intended to re-establish the Fundamental Marxian Theorem by reformulating it in terms of optimum values. Thus, droping assumptions (S3) and (S4), but still maintaining the other assumptions, Morishima considered the following system. Let $\mathbf{A}$ be the matrix of input coefficients, $\mathbf{L}$ the row vector of labor inputs, and $\mathbf{B}$ the matrix of output coefficients, i.e. the ith column of $\mathbf{B}$ is the vector of outputs of production process $\tilde{\mathbf{x}}_i$. Moreover, let $\mathbf{b}$ be the total consumption vector of the working class. According to Morishima, the necessary labor-time, i.e. the labor-time socially necessary to produce the worker's consumption basket $\mathbf{b}$, min $\mathbf{Ly}$, is obtained by solving the following linear program:

(P1) Minimize $\mathbf{Ly}$, with respect to $\mathbf{y}$, subject to

$$\mathbf{By} \geqq \mathbf{Ay} + \mathbf{b}, \qquad \mathbf{y} \geqq 0.$$

Thus, the necessary labor-time is the minimum time required to produce $\mathbf{b}$. The dual of this program is

(P2) Maximize $\Lambda\mathbf{b}$, with respect to Λ, subject to

$$\Lambda\mathbf{B} \leqq \Lambda\mathbf{A} + \mathbf{L}, \qquad \Lambda \geqq 0.$$

If $\mathbf{y}_0$ is a solution to P1 and Λ_0 is a solution to P2, then the Duality Theorem implies that

$$\Lambda_0\mathbf{b} = \mathbf{Ly}_0, \tag{1}$$

and so the components of Λ_0 can be interpreted as some kind of labor-values, which are designated by Morishima as 'optimum values'. Since there may be infinitely different solutions to P2, these optimum values are not uniquely determined, although the value of the consumption vector per worker $\mathbf{c} = N^{-1}\mathbf{b}$ (where N is the number of workers) is the same for every optimum value vector. For let T be the length of the working day and suppose that the workers do not save. Then the exploitation rate is defined as

$$\varepsilon = \frac{TN - \mathbf{Ly}_0}{\mathbf{Ly}_0}.$$

Since $\mathbf{Ly}_0 = \min \mathbf{Ly}$ is uniquely determined (even though $\mathbf{y}_0$ is not), equation 1 implies that

$$\varepsilon = \frac{TN - \Lambda_0\mathbf{c}}{\Lambda_0\mathbf{c}}$$

is the same for every solution Λ_0. As Morishima put it:

> Like actual values, optimum values may not be unique if joint outputs and alternative methods of production are admitted. But unlike actual values, they give a unique evaluation of $\mathbf{c}$; that is to say, $\Lambda_0\mathbf{c}$ takes on the same value for all optimum value systems.[31]

In a similar fashion, although in a more general technology, Roemer (1980, 1981), defined the concept of labor-value as the solution to a programming problem. Instead of a Von Neumann technology, Roemer considered a closed convex[32] set Y containing the null vector $\mathbf{0}$ in which every bundle of commodities can be produced and every positive output requires labor in order to be produced. Assuming (S2), that labor is homogeneous, Roemer defines the labor-value of a bundle $\mathbf{y}$ as the number

$$\min\{\mathbf{x} : [\mathbf{x}, \underline{\mathbf{x}}, \overline{\mathbf{x}}] \in Y \text{ and } \overline{\mathbf{x}} - \underline{\mathbf{x}} \geqq \mathbf{y}\}.$$

Clearly, this number is not necessarily a solution to a linear programming problem, because the set Y is described in very general terms. Like Morishima's definition of labor-values, however, Roemer's definition also allows a unique determination of the exploitation rate for every labor process in Y.

Roemer's definition of labor-value, which clearly encompasses that of Morishima as a special case, has been criticized by some

Marxian economists. The standard criticism is that the labor-value of a bundle of commodities should be determined by the average techniques being actually used in the economy, not by those which are the most efficient in the utilization of labor. According to Roemer, this objection

> is a mainly semantic matter. By defining the labor-value of y as we have done, we are asking for the labor-efficient way of producing y, using the aggregate production set Y. If a "socially average" technique is inferior to this, then we would be injecting some sort of inefficiency into our conception of labor-values, which is not in the Marxian spirit.[33]

At any rate, the main problem of his definition of labor-value is not so much that it characterizes value in terms of labor-efficient processes, but rather that it characterizes value in such a way that it presupposes that labor is homogeneous.

Independently of his definition of labor-value, Roemer has provided a new and important theoretical framework for MTV, within the convexity assumptions which are nowadays usual in mathematical economics, and proved outstanding results regarding the existence of Marxian reproducible equilibria in a variety of interesting models. These results stand, even if Roemer's definition of labor-value is rejected, and will provide the framework for the formulation of the axiomatic foundations of MTV in the present book. I do not claim that this framework is absolute: As I said, it depends on the convexity assumptions which contemporary mathematical economics takes for granted, but these assumptions could be eliminated as the science of economics progresses.[34] As a matter of fact, we shall maintain assumption (S5), which is a blend of convexity and linearity.

Yet, the main reason to adopt Roemer's framework is not the fact that he adopts the standard terms and assumptions of contemporary mathematical economics, but rather the fact that Roemer *does* consider the behavior of the firm, as well (in some cases) the existence of financial capital markets. This is required to formulate the market conditions that produce a uniform profit rate. The task of a Marxian theoretician would be to show that even under such equilibrium-productive behavior, prices are regulated by socially necessary labor-time.

Thus, in view of what has been previously said, the first foundational problem that I want to formulate here is this: *Is it possible to generalize Roemer's theoretical framework in order to get rid of* (S2)? In mathematical terms, what the elimination of (S2) means within Roemer's theoretical framework is that the labor-power vector **x** of any production process in Y is not necessarily unidimensional, i.e. a real number, but in the general case it is an n vector for $n \geq 1$. Obviously, in such a case the definition of labor-value, as the minimum amount of labor required to produce a bundle of commodities as a net output, breaks down. How to define the concept of labor-value within such a generalized structure?

Before addressing this question, it will be convenient to introduce the methodological tools that shall be used. The next two chapters are devoted to this. In the fifth chapter we will address this question. The remaining assumptions (S6)-(S8) will be discussed in subsequent chapters.

Chapter 3

STRUCTURES AND REPRESENTATIONS

The present chapter is devoted to introduce the main metatheoretical concepts that shall be used to attack the foundational problems of MTV. Basically, they are two, namely, the concept of a structure and that of a representation. Roughly, a structure is an array of sets, and of relations over such sets, subject to certain conditions and laws. Notice that, in this sense, the usual algebraic structures are indeed structures. I will introduce this concept of structure in the first section of the present chapter. In some cases, structures can be used to represent real situations. As a matter of fact, the structures that we shall consider in subsequent chapters intend to represent market capitalist economies or some of their aspects. The structures used in the sciences may or may not contain numerical functions (like mass, force, temperature, and the like). Some of them include qualitative relations —like the abstract labor relation—, and when that is the case sometimes it arises the problem whether it is possible to *measure* such relations by means of numerical functions that somehow "mirror" the structure of the relation in some appropriate mathematical structure. Such a function —when it exists— is called a *representation* of the relation. I shall devote the second section of this chapter to discuss the concept of representation.

3.1 STRUCTURES

In order to motivate a general definition of structure, let us start by considering structures of first order languages. A *first order logic* is a family of symbols divided into categories (a first order language), together with certain rules of formation, transformation and interpretation. The symbols of a first order language are mainly divided into logical and non-logical symbols. The *logical symbols* of the language are

$$), (, \neg, \rightarrow$$

and the countable sequence of *individual variables*

$$x_1, x_2, x_3, ..., x_k, ... \quad .$$

) and (are called *grouping symbols* or *parentheses*; $\neg$, $\rightarrow$ are the *connectives*. It is optional to include among the logical symbols the identity symbol '=', which is considered as a (constant) two-place predicate. The *parameters* consist of the universal quantifier $\forall$ and, for each positive integer n, of a set (possibly empty) of symbols called *n-place predicate symbols*; of a set (possibly empty) of symbols called *constant symbols*; and, for each positive integer n, of a set (possibly empty) of symbols called *n-place function symbols* (a minimal language contains at least one predicate symbol). The remaining usual connectives

$$\wedge, \vee, \leftrightarrow,$$

as well as the existential quantifier $\exists$, can be defined in terms of the previous logical symbols and the universal quantifier. This is done below.

An *expression* is any finite sequence of symbols. The formation rules stipulate how to build expressions out of given expressions. There are two kinds of expressions: terms and formulas. These concepts are defined recursively by means of certain operations. In order to define the concept of a term, for each n-place function symbol f the *term-building operation* F_f is needed, that takes expressions $\varepsilon_1, ..., \varepsilon_n$ and the n-place function symbol f to build the expression $f\varepsilon_1 \cdots \varepsilon_n$. In terms of the term-building operations, the set of *terms* is defined as the smallest set that contains the individual variables and the constant symbols, and which is closed under the term-building operations. Thus, if 'c_1', 'c_2' are constant symbols 'f' is a one-place function symbol and 'g' is a two-place function symbol, then examples of terms are 'x_{54}', 'c_1', 'c_2', 'fc_1' and '$gx_{54}fc_1$'. Analogously, in order to define the concept of a formula, the *formula-building operations* $F_\neg, F_\rightarrow, U_i$ are required. These operations yield the following results, for arbitrary formulas ϕ, ψ and positive integer i:

$$F_\neg(\phi) = (\neg\phi)$$
$$F_\rightarrow(\phi, \psi) = (\phi \rightarrow \psi)$$
$$U_i(\phi) = \forall x_i \phi$$

An *atomic formula* is an expression of the form

$$Pt_1 \cdots t_n$$

where P is an n-place predicate and $t_1, ..., t_n$ are terms. In terms of the former operations, the set of well-formed formulas (wff) can be defined as the smallest set containing the atomic formulas which is closed under the formula-building operations. The remaining connectives and the existential quantifier are defined by the following conditions. For arbitrary formulas ϕ, ψ, and variable x_i:

$(\phi \wedge \psi)$ abbreviates $\neg(\phi \rightarrow \neg\psi)$

$(\phi \vee \psi)$ abbreviates $((\neg\phi) \rightarrow \psi)$

$(\phi \leftrightarrow \psi)$ abbreviates $((\phi \rightarrow \psi) \wedge (\psi \rightarrow \phi))$

$\exists x_i \phi$ abbreviates $(\neg(\forall x_i(\neg\phi)))$.

A variable x_i may be free in a formula. This concept is defined recursively as follows:[1]

(1) For atomic ϕ, x_i occurs free in ϕ iff x_i occurs in (is a symbol of) ϕ

(2) x_i occurs free in $\neg\phi$ iff x_i occurs free in ϕ

(3) x_i occurs free in $(\phi \rightarrow \psi)$ iff x_i occurs free in ϕ or ψ

(4) x_i occurs free in $\forall x_j \phi$ iff x_i occurs free in ϕ and $x_i \neq x_j$.

A *sentence* is a wff in which no variable occurs free.

The transformation rules stipulate which formulas can be inferred from given formulas. There are many possible sets of transformation rules but it is possible to select a rather large set of logical axioms and only *modus ponendo ponens* as an inference rule. In such a case, if Λ is the set of logical axioms of the system, ϕ any wff, and Γ a set of wff, a *deduction of ϕ from Γ* is a sequence $\langle \alpha_0, ..., \alpha_n \rangle$ of wff such that $\alpha_n = \phi$ and for each $i \leq n$ either

(1) α_i is in $\Gamma \cup \Lambda$, or

(2) for some j and k less than i, α_i is obtained by modus ponens from α_j and α_k, where $\alpha_k = \alpha_j \rightarrow \alpha_i$.

A formula ϕ is said to be a *theorem of Γ* (in symbols: $\Gamma \vdash \phi$) iff there is a deduction of ϕ from Γ. For a particular selection of the set Λ of logical axioms, the reader is referred to Enderton (1972).[2]

The rules of interpretation for a first order language are clustered around the concept of a structure for the same language. A *structure*[3] $\mathfrak{A}$ for a first order language is a function from the set of parameters into some family of sets that satisfies the following conditions:

(1) $\mathfrak{A}$ assigns to the quantifier $\forall$ a nonempty set $|\mathfrak{A}|$, called the *universe* of $\mathfrak{A}$

(2) $\mathfrak{A}$ assigns to each n-place predicate symbol P an n-ary relation $P^{\mathfrak{A}} \subseteq |\mathfrak{A}|^n$; i.e. $P^{\mathfrak{A}}$ is a set of n-tuples of members of the universe

(3) $\mathfrak{A}$ assigns to each constant symbol c a member $c^{\mathfrak{A}}$ of the universe $|\mathfrak{A}|$

(4) $\mathfrak{A}$ assigns to each n-place function symbol f an n-ary operation $f^{\mathfrak{A}}$ on $|\mathfrak{A}|$; i.e. $f^{\mathfrak{A}}: |\mathfrak{A}|^n \rightarrow |\mathfrak{A}|$.

Given a sentence ϕ in a language, and a structure $\mathfrak{A}$ for the language, it is of the utmost importance to ask whether ϕ is true in $\mathfrak{A}$. To give a precise meaning to this question, the concept of truth needs to be properly defined. In order to deal with the occurrence of variables —which may be free— in the formulas, the set of all functions s from the set of variables into the universe $|\mathfrak{A}|$ is introduced. Each of these functions s assigns an element of the universe to each variable and can be extended to a function $\bar{s}$, which is defined by recursion as follows:

(1) For each variable x_i: $\bar{s}(x_i) = s(x_i)$

(2) For each constant symbol c: $\bar{s}(c) = c^{\mathfrak{A}}$

(3) If $t_1, ..., t_n$ are terms and f is an n-place symbol, then

$$\bar{s}(f\,t_1 \cdots t_n) = f^{\mathfrak{A}}(\bar{s}(t_1), ..., \bar{s}(t_n)).$$

If s is one of the former functions, $s(x_i|d)$ denotes the function which is identical to s except that it assigns the element d of the universe to x_i.

Using the functions s and their extensions $\bar{s}$, we shall define the notion of a formula ϕ being satisfied by $\mathfrak{A}$ with s (in symbols: $\vDash_{\mathfrak{A}} \phi\,[s]$). This is done as follows. For arbitrary formulas ϕ, ψ and variable x_i:

(1) $\vDash_{\mathfrak{A}} t_1 = t_2\,[s]$ iff $\bar{s}(t_1) = \bar{s}(t_2)$

(2) If P is an n-place predicate symbol,

$$\vDash_{\mathfrak{A}} P t_1 \cdots t_n\,[s] \quad \text{iff} \quad \langle \bar{s}(t_1), ..., \bar{s}(t_n) \rangle \in P^{\mathfrak{A}}$$

(3) $\vDash_{\mathfrak{A}} \neg\phi\,[s]$ iff $\nvDash_{\mathfrak{A}} \phi\,[s]$

(4) $\vDash_{\mathfrak{A}} (\phi \rightarrow \psi)\,[s]$ iff either $\nvDash_{\mathfrak{A}} \phi\,[s]$ or $\vDash_{\mathfrak{A}} \psi\,[s]$ or both

(5) $\vDash_{\mathfrak{A}} \forall x_i \phi\,[s]$ iff for every $d \in |\mathfrak{A}|$, it is the case that

$$\vDash_{\mathfrak{A}} \phi\,[s(x|d)].$$

Naturally, a wff ϕ is said to be *satisfiable* iff there is a structure $\mathfrak{A}$ and a function s such that $\vDash_{\mathfrak{A}} \phi\,[s]$. A set of formulas Γ is *satisfiable* iff there is a structure $\mathfrak{A}$ and a function s such that $\mathfrak{A}$ satisfies every member of Γ with s. The crucial concept of logical implication can be defined now. Let Γ be a set of wffs and ϕ a wff. Γ is said to *logically imply* ϕ, or ϕ to be a *logical consequence* of Γ ($\Gamma \vDash \phi$) iff, for every structure $\mathfrak{A}$ for the language and every function s from the set of variables into the universe of $\mathfrak{A}$: if $\mathfrak{A}$ satisfies every formula in Γ with s, then $\mathfrak{A}$ also satisfies ϕ with s.

It can be shown that if ϕ is a sentence and $\mathfrak{A}$ any structure, then either ϕ is satisfied by $\mathfrak{A}$ with every s, or not at all. In the first case we say that ϕ is *true* in $\mathfrak{A}$, or that $\mathfrak{A}$ is a *model* of ϕ. Thus, in particular, if ϕ is a sentence and Γ a set of sentences, then ϕ is a logical consequence of Γ iff every model of Γ is a model of ϕ.

I have just described a first order logic in a succint way. The most important result within first order logic is Gödel's Completeness Theorem, according to which every logical consequence of a set Γ of formulas is also deducible from Γ. This result implies the Compactness Theorem (also known as Finiteness Lemma). The Compactness Theorem asserts that if Γ is a set of formulas such that every finite subset of Γ is satisfiable, then Γ is also satisfiable. Using this theorem, Robinson (1961) proved the existence of a proper extension of the real number system that contains both infinitely large and infinitely small (yet nonzero) numbers. The extreme formal rigour and precision of first order languages, and of their semantic, is required to obtain results like this. Indeed, many important aspects of structures can be studied by means of first order methods.

Given a structure $\mathfrak{A}$ for a first order language, we define the *theory of* $\mathfrak{A}$ (Th $\mathfrak{A}$) as the set of all the sentences of the language which are true in $\mathfrak{A}$. Analogously, if $\mathfrak{S}$ is a class of structures, the *theory* of $\mathfrak{S}$ (Th $\mathfrak{S}$) is the set of all sentences true in each of the elements of $\mathfrak{S}$. A *theory* is defined as a set of sentences of the language which is closed under logical implication, i.e. T is a *theory* iff

$$T \vDash \phi \quad \Rightarrow \quad \phi \in T.$$

Let Σ be a consistent set of sentences and denote by Mod Σ the set of all models of Σ. The set of *consequences* of Σ (Cn Σ) is defined as Th Mod Σ. It is easy to see that a set of sentences T is a theory iff $T = \text{Cn } T$.

Roughly speaking, a set Σ of sentences is *decidable* if there is an algorithm which can tell, in a finite number of steps, whether a given sentence is or is not an element of Σ.[4] A theory T is *axiomatizable* iff there is a decidable set of sentences Σ such that $T = \text{Cn } \Sigma$. In particular, if Σ is finite, then T is said to be *finitely axiomatizable*.

Some philosophers of science attempted to identify the scientific theories as sets of sentences that could be translated into first order sentences and put together in a consistent set T; furthermore, they also claimed that T was axiomatizable, so that the scientific theories could actually be identified with a decidable set Σ, a set of axioms for T. These assumptions are the basis of the so-called statement view of theories. The statement view of theories is misleading because it suggests that the structures usually found in the sciences are structures for first order languages. Unfortunately (because first order model theory is beautiful and contains very deep, useful results) many structures in science are not structures for a first order language. As examples, consider the topological and the probability spaces. Recall that a topological space is a pair $\mathfrak{X} = \langle X, \tau \rangle$ such that X is a nonempty set and τ is a family of subsets of X containing the empty set, the set X itself, and closed under finite intersections and arbitrary unions. It is easy to see that the topological spaces are not models of a first order theory (i.e. of the language in which the sentences of such theory are formulated). A first attempt to provide a first order language for which the topological spaces would be the structures would have to start by noticing that the universe of a structure for any such language would have to be the power set $\wp(X)$ of X, since the theory needs to be able to refer to all the subsets of X and, moreover, τ needs to be thought of as a one-place predicate (i.e. a subset of $\wp(X)$), $\cap$ needs to be thought of as a binary operation over $\wp(X)$, and $\emptyset$ and X have to be considered as constants, i.e. as elements of the universe $\wp(X)$. From this point of view, the first two axioms of topology can be formulated as follows:

$\emptyset \in \tau \wedge X \in \tau$

$\forall x \forall y ((x \in \tau \wedge y \in \tau) \to x \cap y \in \tau).$

The problem is —as the perspicacious reader must have already noticed— that there is no way of formulating the third axiom, because $\cup$ would have to be treated as an infinitary operation over $\wp(X)$. The problem remains even if we move onto second order languages, because the problem is not the lack of quantification over predicate or function variables, but rather the need to treat $\cup$ as an operation over the universe of the structure.

An apparent way out is to treat $\cup$ (and perhaps also $\cap$) as unary operations over $\wp(\wp(X))$, i.e. if $Y \subseteq \wp(X)$, then $\bigcup Y \in \wp(X)$ and $\bigcap Y \in \wp(X)$. But then the new problem arises that in such a case we would have to consider $\wp(\wp(X))$ as the universe (since the operations in first order logic are treated as functions having some Cartesian power of the universe as domain), in which case the operations would not be closed in the universe. Another shortcoming of this approach is that the "individuals" would be families of subsets of X; thus, in particular the topology τ would be an individual, but then the objects of level "lower" that τ —like the elements of τ— cannot be referred to in any way whatsoever and so none of the fundamental axioms can be formulated. Similar problems arise when a first order formulation of the Kolmogorov axioms for probability spaces is attempted.

The root of the former problems is that many theories consider simultaneously several ranks of objects, i.e. roughly speaking, individuals, sets of individuals, sets of sets of individuals, or even more complex constructions, whereas first order structures involve only individuals, finitary operations over individuals and relations over individuals.

The notion of a rank of objects can be given a precise definition. Consider an arbitrary nonempty set X and define

$$X_0 = X; \quad X_{n+1} = \wp\left(\bigcup_{i=0}^{n} X_i\right).$$

The *superstructure* over X is the set $\bigcup_{i<\omega} X_i$ and is denoted by $\check{X}$. An object $x \in \check{X}$ is said to be of *rank* $k > 0$ iff x is an element of X_k and not an element of X_j for $j < k$ (if $x \in X_k$, then $x \in X_m$ for $m > k$). We shall introduce a concept of structure that is able to handle theoretical structures having any finite number of objects of arbitrary ranks. In order to do so, notice that $\check{X}$ can be seen as a model of Zermelo-Fraenkel theory of sets. The language of this theory contains the two-place predicate symbol '$\in$' as unique primitive. The axioms of this theory[5] guarantee the existence of the sets $\bigcup y$ and $\bigcap y$ for any nonempty set $y \in \check{X}$. Thus $\bigcup$ and $\bigcap$ can be seen as (definable) unary operations over $\check{X}$. Taking advantage of this fact the concept of a topological space can now be defined as as a pair $\langle X, \tau \rangle$ of entities in $\check{X}$ that satisfies the first order sentences:

(T1) $\emptyset \in \tau \wedge X \in \tau$

(T2) $\forall x((x \subseteq \tau \wedge \mathit{finite}(x)) \rightarrow \bigcap x \in \tau)$

(T3) $\forall x(x \subseteq \tau \rightarrow \bigcup x \in \tau)$.

(Indeed, the predicate *'finite'* is definable within ZF).

   This first order axiomatization of topology suggests that any structure can be axiomatically characterized within a first order language, provided that the generator X of the superstructure $\check{X}$ is suitably chosen. This can be done even when the structure under consideration is composed by sets as dissimilar as one of physical things and another of numbers. For instance, suppose that we have a structure $\langle A, \mathbf{R}, m \rangle$, where A is a set of physical bodies, and m a function $m: A \rightarrow \mathbf{R}$ assigning to each element of A a positive real number (say, its mass). The minimum superstructure required to build this structure is the one having as generator the set $X = A \cup \mathbf{R}$. Sentences with restricted quantifications of the form 'for every $x \in A...$' can be reformulated within the language as sentences of the form

$$\forall x(P(x) \wedge \, ... \, ,$$

provided that a predicate 'P', to be interpreted as the set A, is introduced into the language.

The former discussion suggests a general notion of structure which is appropriate for the purposes of axiomatizing scientific theories. Roughly, such a structure is nothing but a finite array of sets $D_1, ..., D_k$, and of set-theoretic constructions $R_1, ..., R_m$ over such sets, that satisfies certain sentences formulated within an extension of the language of ZF. For instance, a topological space is a structure in this sense, where $D_k = X$ and $R_m = \tau$, that satisfies the sentences T1-T3 above. The sentences a structure satisfies are indeed first order sentences, but this does not mean that the structure is a structure for the language in which these sentences are formulated. Rather, they are elements of a structure $\langle \check{X}, \in \rangle$ for the same language. This prevents us from exploiting certain results of first order model theory, but at least we can refer to all the entities involved in a very natural way and, furthermore, we can freely employ all the results of ZF for our purposes. Since, in particular, classical mathematics can be developed within the framework of ZF, it follows that we have all of classical mathematics at our disposal. This is not the place to develop the mathematical concepts that are being used in the reconstruction of MTV; the interested reader can find a detailed development of the real number system in Suppes (1972) and Landau (1966). The concepts from linear algebra used here, like those of vector and linear space, are developed for instance in McLane and Birkhoff (1967).

This concept of structure can be given a very precise formulation. What it is required to do is to identify a whole class of structures in the most general terms. For instance, topology needs to identify the class of all the topological spaces, probability theory that of all probability spaces, and so on. This is done through the concept of a structure species, which was introduced originally by Bourbaki (1968), but better exposed by Balzer, Moulines and Sneed (1987). My own exposition will follow closely this last one. The definition of structure species is the last of a series starting with the definition of a k-type.

For each positive integer k, the concept of a k-type σ is defined inductively as follows:

(1) for each positive integer $i \leq k$: i is a k-type

(2) if σ is a k-type then $\wp(\sigma)$ is a k-type

(3) if σ_1 and σ_2 are k-types then $(\sigma_1 \times \sigma_2)$ is a k-type.

From an ontological point of view, k-types are set-theoretic entities build up from positive integers (which themselves are sets). But we shall identify each k-type σ with an operation assigning to certain sets a set-theoretic construction based on those sets. These sets have to be k in number and shall be denoted by $D_1, ..., D_k$; the construction that the k-type σ associates to these sets will be called the 'echelon set of type σ' over $D_1, ..., D_k$ and denoted by $\sigma(D_1, ..., D_k)$. The definition of this concept is also inductive and proceeds as follows:

(1) if σ is some i $(i \leq k)$ then $\sigma(D_1, ..., D_k) = D_i$

(2) If σ_1 is a k-type and σ has the form $\wp(\sigma_1)$, then

$$\sigma(D_1, ..., D_k) = \wp(\sigma_1(D_1, ..., D_k))$$

(3) If σ_1 and σ_2 are k-types, and σ has the form $(\sigma_1 \times \sigma_2)$ then

$$\sigma(D_1, ..., D_k) = \sigma_1(D_1, ..., D_k) \times \sigma_2(D_1, ..., D_k).$$

As it is usual in set-theory, we may identify the objects of set theory with their own names and think also of the k-types as term-building operations. Consider an extension of the language of ZF containing all the terms generated from the terms $D_1, ..., D_k$ by such operations. If 'c' is a constant symbol of the language denoting a set, then we shall call any formula of the language of the form

$$c \in \sigma(D_1, ..., D_k)$$

a *typification*.

It is usual to find in scientific structures a distinction between base sets and auxiliary sets. Base sets are those that contain the empirical objects the theory deals with, whereas auxiliary sets are mathematical sets used to represent magnitudes of objects pertaining to base sets. That is why in the definition of a type it is convenient to talk of $(k + m)$-types, where k is the number of base sets and m is the number of auxiliary sets. Thus, a *type* τ is an array $\langle k, m, \sigma_1, ..., \sigma_n \rangle$ such that

(1) k is a positive integer and m is a nonnegative integer

(2) $\sigma_1, ..., \sigma_n$ are $(k + m)$-types.

If $\tau = \langle k, m, \sigma_1, ..., \sigma_n \rangle$ is a type, a *structure of type* τ is a tuple $\langle D_1, ..., D_k, A_1, ..., A_m, R_1, ..., R_n \rangle$ such that $D_1, ..., A_m$ are sets and, for each $i \le n$, the typification

$$R_i \in \sigma_i(D_1, ..., D_k, A_1, ..., A_m)$$

is true in any model of ZF generated from a set containing $(\bigcup_{j=1}^{k} D_j) \cup (\bigcup_{l=1}^{m} A_l)$.

Let $\tau = \langle k, m, \sigma_1, ..., \sigma_n \rangle$ be a type, and

$$\mathfrak{D} = \langle D_1, ..., D_k, A_1, ..., A_m, R_1, ..., R_n \rangle$$

a structure of type τ. Consider an extension of the language of ZF containing constant symbols for the sets

$$D_1, ..., D_k, A_1, ..., A_m, R_1, ..., R_n.$$

A formula of such language containing no ocurrences of constant symbols other than symbols among these is said to *apply to* $\mathfrak{D}$. If τ is a type as before, a *structure species of type* τ is a tuple $\rho = \langle \tau, \phi_1, ..., \phi_s \rangle$ such that, for all $i \le s$, ϕ_i applies to some structure of type τ. Naturally, a *structure species* is a structure species of some type τ. Among the formulas ϕ_i ($1 \le i \le s$) that apply to some structure of type τ, some contain occurrences of symbols for the base sets $D_1, ..., D_k$ and for precisely *one* of the relations $R_1, ..., R_n$. Formulas of this type characterize these relations and so they will be called *characterizations*; notice that, in particular, typifications are characterizations. Scientific laws are not characterizations because the least that can be said about them is that they establish connections between different relations.

We have thus reached the definition of the concept we were seeking. We shall identify a scientific structure precisely with a structure of some species ρ. If $\rho = \langle \tau, \phi_1, ..., \phi_s \rangle$ is a structure species of type $\tau = \langle k, m, \sigma_1, ..., \sigma_n \rangle$, then a *structure of species* ρ is a tuple

$$\langle D_1, ..., D_k, A_1, ..., A_m, R_1, ..., R_n \rangle,$$

where $D_1, ..., A_m$ are sets, that satisfies certain formulas $\phi_1, ..., \phi_s$ that apply to it. The relevant concept of satisfaction is the following. The structure

$$\mathfrak{D} = \langle D_1, ..., D_k, A_1, ..., A_m, R_1, ..., R_n \rangle$$

satisfies the formula ϕ iff ϕ applies to $\mathfrak{D}$ and ϕ is satisfied in any model of ZF generated from a set that contains $(\bigcup_{j=1}^{k} D_j) \cup (\bigcup_{l=1}^{m} A_l)$. By the *theory of* $\mathfrak{D}$ we understand the set of all sentences that apply to $\mathfrak{D}$ which are true in $\mathfrak{D}$, i.e. which are satisfied by $\mathfrak{D}$. We say that the theory of $\mathfrak{D}$ is *axiomatizable* if there is a decidable set Σ of sentences that apply to $\mathfrak{D}$ such that every element in the theory of $\mathfrak{D}$ is a logical consequence of Σ. Let Γ be a set of sentences. A *model* of Γ is any structure $\mathfrak{D}$ such that the sentences in Γ apply to $\mathfrak{D}$ and are satisfied by $\mathfrak{D}$.

From an ontological point of view, structures are abstract entities (*entia rationis*, as the scholastics would say) that can be object of mathematical research. Some of these entities, in addition, represent real things, systems or processes. Representation theory is a philosophical discipline that aims to gain insight into the way structures are carriers of knowledge about aspects or parts of the real world. The main problem of representation theory is to give an account of the way mathematics applies to reality. The present chapter is quite far from pretending to provide a complete account of the problems and results of that discipline: Its aim is just to sketch how such a discipline would look like and to introduce the concept of representation. It will be useful for this purpose to discuss an important useful case of representation.

3.2 REPRESENTATIONS
3.2.1 THE ONTOLOGICAL FRAMEWORK

According to cosmology and natural history, the Earth, the planets and millions of stars were already quite ancient when *homo sapiens sapiens* began to marvel at the surrounding world. What this means is that the human mind occupies a rather humble place in creation, far away from the place of creator of the natural world. On the contrary, man has always been trying to dominate the natural phenomena for his own ends, a task whose success is always limited by the vastness of the world. Since the times when men were afraid of thunder and ray, that task has been accomplished through an increasing knowledge of the natural phenomena. Originally, this knowledge was rather empirical and unsystematic, but since the rise of astronomy in Babilon, and especially in classical Greece with Eudoxus, that knowledge has become theoretical and systematic. After the scientific revolution in the seventeenth and eighteenth centuries, many more scientific disciplines have arisen, including disciplines that deal with social phenomena as well. Among other things, these disciplines are constituted by conceptual structures and theories by means of which the scientists —as before— intend to obtain knowledge of some parts or aspects of the real world.

The development of mathematical logic in the current century —a discipline that may be divided into model theory, proof theory, recursion theory and set theory— has provided new tools to analyze the conceptual structures and theories produced by the different sciences, giving a new shift to the theory of science, a discipline inaugurated by the Bohemian philosopher and mathematician Bernhard Bolzano with his *Wissenschaftslehre* in 1837. From the point of view advocated here, the theory of science can be defined as the discipline in charge of determining the logical structure and foundations of scientific theories, as well as the way these theories connect to the real phenomena with which they deal. This definition of the theory of science presupposes that there is a way of referring to the real world which is independent of, and in some sense "previous" to the language of any given scientific theory. This presupposition is true: That way is provided by the conceptual system and language of the *philosophia de ente*, enriched with proper and common names referring to natural kinds, substances and accidents of substances. By means of such a language we certainly can refer to the *entia* with which the sciences deal, and formulate propositions which are true about such *entia*, in a way independent of the technical language of these sciences. The *philosophia de ente* contains the conceptual apparatus required to discuss, for example, whether an object considered by a theoretical model is one *ens*, or merely seems to have individual unity just because of the way in which the mind considers it; or whether an object denoted by the language of a scientific discipline is really or merely conceptually existing —if it is an idealization.

Someone might object that if we had an independent way of referring to the *entia* with which the sciences deal, then the mathematical structures representing them would be redundant and useless. The reply to this is that one thing is to refer to something and quite another to represent it mathematically. The reason to pursue a mathematical representation of a real being, process or system is mainly to probe deeper into its proper operation, relations or intrinsic nature, and quite often to obtain *measurements* of some quantity inhering in it, that is needed for practical or theoretical purposes.

The *philosophia de ente* I have in mind —also known as ontology or metaphysics— was initiated by Aristotle some five centuries BC and perfected by the scholastics; in a very outstanding and systematic way by Francis Suarez. The *philosophia de ente* is previous to any science in the sense that its language and conceptual apparatus do not presuppose that of any science. This is due to two facts: (1) Its subject matter is being, the most general topic of all; and (2) being is what is first given to the understanding. I take this last sentence to mean that the basic structure and unsystematic comprehension of being are given in an experience that does not need to be scientific. For instance, the distinction substance-accident, or the distinction quality-relation, can be and in fact were established independently of every scientific theory. They are the result of a theoretical elaboration of experiences mainly obtained in the transformation of nature by man through labor. This does not mean that there cannot be any feedback from science to ontology. Clearly, the concept of a system, or that of a relation non-reducible to qualities, are examples of ontological concepts, not belonging to the original *corpus* of the *philosophia de ente*, strongly demanded by contemporary science. But ontology cannot be reduced to any science and has to be cultivated with its own methods and procedures.

Hegel referred to ontology in his lesser *Logic* as part of that discipline that he labeled 'Metaphysic of the Past'. Yet, what Hegel had in mind here was mainly Christian Wolf's metaphysics, which included ontology as one of its branches, the others being cosmology, rational psychology and natural theology. Hegel claimed that this metaphysics —which he also characterized as a metaphysics of the understanding (*Verstände*)— was doomed to failure in its treatment of "infinite objects", i.e. God, the Soul and the World, due to the fact that it intended to treat these objects with the finite categories of understanding. Yet, Hegel made no objection to its treatment of finite being:

> In finite things it is no doubt the case that they have to be characterized through finite predicates: and with these things the understanding finds proper scope for its special action. Itself finite, it knows only the nature of the finite. Thus, when I call some action a theft, I have characterized the action in its essential facts; and such a knowledge is sufficient for the judge. Similarly, finite things stand to each other as cause and effect, force and exercise, and when they are apprehended in these categories, they are known in their finitude. But the objects of reason cannot be defined by these finite predicates. To try to do so was the defect of the old metaphysic.[6]

I think that the defects that both Kant and Hegel saw (in different ways) in Wolf's metaphysics (in his ontology, in particular) cannot be attributed without any further considerations to Suarez and the previous scholastics,[7] but leaving aside the problem whether Hegel's criticism of Wolfian ontology can be extended to scholastic ontology (which I think is a very interesting problem), the former remarks about finite being surely apply quite well to the Aristotelian-scholastic treatment of finite *entia*, i.e. to what some philosophers would refer to as the "ontic" aspects of reality. It is important to make these remarks here because the ontological framework that was behind Marx's construction of *Capital* was precisely Hegel's logic. I shall address the problem of characterizing Marx's dialectical method below, as well as his relation to Hegel's metaphysics, but it is important to stress at this point that the theory of measurement I am advocating makes use of rather general characterizations of finite *entia* in terms of categories of the "understanding", as a "moment" in the whole process of theoretization.[8]

The guiding line of our research shall be the problem of the application of mathematics to reality. The concept by means of which I will attempt to explicate this application is that of fundamental measurement. In order to introduce this concept, we shall define an *ontological structure* as a structure of some species, of the form $\langle A, R_1, ..., R_n \rangle$, that satisfies certain ontological sentences $\phi_1, ..., \phi_s$, such that the elements of A are real *entia* and the R_i $(i = 1, ..., n)$ are set-theoretic relations among the elements of A that represent real relations or operations among the same elements. If r is a real relation among substances, we say that R *represents* r iff R is a set of tuples of substances related by r. More precisely, let '$Fx_1 \cdots x_k$' express the fact that the *entia* $x_1, ..., x_k$ are related by r. Then we say that R represents r iff

$$\langle x_1, ..., x_k \rangle \in R \quad \text{implies} \quad Fx_1 \cdots x_k.$$

A *numerical structure* is a structure like the previous one, except that the underlying set A is a set of real numbers. A *fundamental measurement* of the real quantities, operations or relations $r_1, ..., r_n$ of or among *entia* $x_1, ..., x_m$ is a homomorfism φ from the structure $\mathfrak{A} = \langle A, R_1, ...R_n \rangle$ into a numerical structure $\mathfrak{B} = \langle B, S_1, ..., S_n \rangle$, where $A = \{x_1, ..., x_n\}$ and R_i *represents* r_i $(1 \leq i \leq n)$. When such a homomorphism exists, we say that $\mathfrak{B}$

represents $\mathfrak{A}$, and also that φ *represents* the relations R_i. Thus, the function $\varphi{:}A \rightarrow B$ represents the relation R_i (and so, indirectly, also the relation r_i) iff for every $\langle x_1, ..., x_n \rangle \in R_i$:

$$\langle x_1, ..., x_n \rangle \in R_i \Leftrightarrow \langle \varphi(x_1), ..., \varphi(x_n) \rangle \in S_i.$$

A *representation theorem* for an ontological structure $\mathfrak{A}$ is a statement asserting the existence of a function representing the relations of $\mathfrak{A}$, and also establishing up to what point is that function unique, that can be derived logically from the ontological sentences $\phi_1, ..., \phi_s$. The clause of the representation theorem asserting the existence of the representation is called the *existence part*; the one asserting the degree of uniqueness is the *uniqueness part*.

The role of ontology in the establishment of a representation theorem consists of providing the conceptual apparatus required to discuss and formulate the ontological axioms $\phi_1, ..., \phi_s$. It will be profitable to illustrate this role with an example. This is the contents of the next section.

3.2.2 A CASE OF REPRESENTATION

The situation is the following. There is a material substance —say a wooden beam— having the shape of a parallelepiped. The height of any of these parallelepipeds is the length of any of the segments orthogonal to its bases and enclosed by these bases. According to Francis Suarez, who follows the Philosopher in this respect,[9] these segments are not imaginary, since they are real beings in the category of quantity, inhering in the given beam. Our task is to make ontological sense of the measurement of the lengths of these segments —which we shall call main segments— and their potential subsegments.

Indeed, not any assignment of numbers to the segments would count as a measurement of their lengths. The first condition that a measurement has to fulfill is that it must assign the same number to segments of the same length, and a larger number to the longest segment of any pair of segments. Another requisite is that if a segment can be divided into two segments x, y, then the numbers assigned to the segment and its parts x, y must be such that the sum of the numbers assigned respectively to x, y has to be equal to the number assigned to the whole segment, i.e. the measurement has to be *extensive*. A question that naturally arises is how fine is the main segment grained, what are the smallest segments into which a segment can be divided, or whether the division can continue indefinitely. This raises the old metaphysical problem of the composition of the continuum, a problem which was characterized by Leibniz as one of the two labyrinths of the human mind.[10] This deep metaphysical problem has a direct bearing on the choice of the axioms guaranteeing the existence and uniqueness of the measurement. Krantz *et al.* (1971) introduced a regularity axiom for extensive measurement that can only be interpreted in two ways: The segments can be infinitely divided (this is the Leibnizian view), or there is a smallest subdivision d such that the length of any other subdivision is a multiple of that of d (this is a version of the opposite view). In general, the opposite view is precisely that the segments can be divided into a finite number of smallest parts, their lengths not necessarily being multiples of the smallest part. According to contemporary physics, matter cannot be divided *ad infinitum*, and so it would seem to support the finitistic metaphysical view.

The question that arises now is whether the mereological structure of any segment guarantees the existence of length measurements, assuming that the finitistic view is true. Clearly, since length measurement consists of comparing any length with a common length taken as unit, i.e. in determining "how many concatenated replicas" of this unit are equivalent to any given length, if there is no smallest part, or segment, such that every other segment is a multiple of the smallest part, then no measurement is possible. The only way of measuring the segments into which the height of the segment can be divided is then to bring a unit from outside the segment, such that all those segments are multiples of the given unit. Notice that this presupposes (i) that there is another material substance such that it has a line segment of the type required, and (ii) that the segments into which the beam heights can be divided are *commensurable*.

In the case of the beams, homogeneity considerations make it plausible that the thinnest beam slice has the same width everywhere, and so in this case the regularity axiom holds with the interpretation that there is a smallest subdivision d such that the length of any other subdivision is an integer multiple of that of d. Notice that the regularity axiom we shall introduce is true also under the Leibnizian conception of the continuum. Obviously, the required representation can be constructed just by assigning the number 1 to any smallest subdivision d and the number k to a segment which is equivalent to "k concatenated copies of d". But this is a rather sloppy operationalistic way of describing the construction. A proper philosophical treatment requires the introduction of more precise conceptual tools.

Consider the set X having as elements a main particular segment, orthogonal to the beam bases and determined by these bases, as well as all those potential parts of it, i.e. the segments into which it can be divided, down to the smallest segments. If x, y are any elements of X, we write $x \succsim y$ iff the magnitude of x is greater than or equal to the magnitude of y. As usual, we write $x \succ y$ if $x \succsim y$ but not $y \succsim x$, and $x \sim y$ if both $x \succsim y$ and $y \succsim x$. Notice that these relations are independent of the actual comparison of the segments by any agent, and so they should *not* be conceived operationalistically. We define now a direction on the beam, for instance with respect to the hands, and designate a left and a right on the beam. We introduce now a set Y of pairs $(y, z) \in X \times X$ as follows: The pair (y, z) is in Y iff there is a segment $x \in X$ such that x can be divided into y and z, and y is the subsegment of x to the left of z; in this case we write $x = y \oplus z$. Notice that the fact that we choose to determine the direction of the segment with respect to the hands does not make the direction a subjective entity. A direction is a real ordering among the parts of a body; in the present case, there are two orderings among the parts of the segment, each one determined by an extreme of the beam, which is the first element in the corresponding ordering. In selecting a right and a left in the beam we are only chosing one of these two previously existent orderings. The following definition introduces axioms which are jointly sufficient to prove the required representation theorem. I discuss their meaning and metaphysical justification below.

DEFINITION 1: $\langle X, \succsim, Y, \oplus \rangle$ is an *Aristotelian extensive structure* iff

(A1) $\langle X, \succsim \rangle$ is a weak order;

(A2) *(Congruence)* If both $x \oplus y$, $z \oplus w$ are defined, and $x \sim z$, $y \sim w$, then $x \oplus y \sim z \oplus w$

(A3) *(Dominance)* If $(x, y) \in Y$, $x \succsim z$ and $y \succsim w$, then there are $u, v \in X$ such that $(u, v) \in Y$ and $u \sim z$, $v \sim w$. Moreover, $x \oplus y \succsim u \oplus v$

(A4) *(Decomposition)* If $(x, y) \in Y$ and $x \oplus y \sim z$, then there exist $u, v \in X$ such that $(u, v) \in Y$, $u \sim x$, $v \sim y$ and $z = u \oplus v$

(A5) *(Associativity)* $(x, y) \in Y$ and $(x \oplus y, z) \in Y$ iff $(y, z) \in Y$ and $(x, y \oplus z) \in Y$; and when both conditions hold,

$$(x \oplus y) \oplus z = x \oplus (y \oplus z)$$

(A6) *(Positivity)* If $(x, y) \in Y$, then $x \oplus y \succ x$

(A7) *(Regularity)* If $x \succ y$, then there exist $z, w, u \in X$ such that $z \sim x$, $w \sim y$, $(w, u) \in Y$ and $z \succsim w \oplus u$

(A8) *(Archimedean Axiom)* Every strictly bounded standard sequence is finite. We say that

$$x_1, ..., x_n, ...$$

is a *standard sequence* iff there is a sequence

$$y_1, ..., y_n, ...$$

such that $y_i \sim y_j$ and $(y_i, y_{i+1}) \in Y$ $(i, j = 1, ..., n, ...)$ and, moreover, there is another sequence

$$z_1, ..., z_n, ...$$

defined by $z_1 = y_1$, $z_n = z_1 \oplus z_{n-1}$, with $x_k \sim z_k$; it is *strictly bounded* iff there is an $x \in X$ such that $x \succ x_n$ for all x_n in the sequence.

Axiom (A1) asserts that any two segments in X are compared by the relation $\succeq$, i.e. that of any two $x, y \in X$ either the length of x is greater than or equal to the length of y, or viceversa. Also, that the relation $\succeq$ is transitive. It can be seen that this axiom is metaphysically true.

(A2) asserts the congruence of any two segments whose parts into which they are divided by two are congruent. The axiom is clearly true.

(A3) affirms that if there is a segment divided into two, whose parts repectively dominate two segments, then there is another segment also divided into two, whose parts are respectively congruent to the dominated segments, and which is itself dominated by the original longest segment. A little spatial reflection reveals that this is correct.

(A4) asserts that if a segment is congruent to a segment divisible into two parts, then the first segment can be divided into two parts which are congruent to the parts of the second segment. It is easy to see that the axiom is true.

(A5) is obvious because the order in which a segment division is given is immaterial.

(A6) is also obvious, because the length of a segment is always greater than the length of any of its proper subsegments.

(A6) is the regularity axiom discussed above in connection with the problem of the composition of the continuum. Far from being clearly true —let alone obviously true— it expresses a particular solution to the metaphysical riddle of the nature of the continuum. The axiom is true, however, if the line segments inhering in the bodies are continuous in the sense accepted both by Leibniz and Aristotle (in *Physics*, Book VI, $231^a20\text{-}231^b21$), or if they are composed of a finite number of congruent subsegments as discussed above.

The sense of axiom (A8) is that if we have a sequence of segments, the length of any term in the sequence being greater than the length of the previous term exactly by the same difference as that of any other pair of consecutive terms, and all the terms of the sequence are strictly dominated by the same segment, then the sequence is finite. Roughly speaking, this prevents the existence of infinitesimal lengths, i.e. of congruent lengths such that no finite addition of the same —no matter how large— will ever surpass the length of a given finite segment. The axiom appears to be also true of real lengths.

These metaphysical assertions are sufficient to prove the following theorem.

THEOREM: *Let $\langle X, \succsim, Y, \oplus \rangle$ be an Aristotelian extensive structure. Then there exists a function $\varphi: X \rightarrow \mathbf{R}$ such that, for all $x, y \in X$,*

(i)
$$x \succsim y \quad \text{iff} \quad \varphi(x) \geq \varphi(y)$$

and

(ii)
$$\text{if} \quad (x, y) \in Y, \quad \text{then} \quad \varphi(x \oplus y) = \varphi(x) + \varphi(y)$$

If another function φ' satisfies (i) and (ii), then there exists an $\alpha > 0$ such that for all nonmaximal $x \in X$ $\varphi'(x) = \alpha\varphi(x)$.

The proof of this theorem is involved and requires the construction of another structure, as well as the proof of several lemmas concerning this other structure. In order to construct it, I shall introduce the following definition.

DEFINITION 2: If $\langle X, \succsim, Y, \oplus \rangle$ is an Aristotelian extensive structure, we let Z be the set of all pairs $(x, y) \in X \times X$ such that there are $z, w \in X$ with $z \sim x$, $w \sim y$ and $(z, w) \in Y$. If $(z, w) \in Y$, we say that $z \oplus w$ is *defined*.

The required new structure will be constructed as follows. It shall be the structure $\langle A, \succsim, B, \circ \rangle$ such that A is the quotient set $X/\sim$ of X with respect to $\sim$, $\succsim$ is the relation given by

$$[x] \succsim [y] \quad \text{iff} \quad x \succsim y,$$

B is the set $\{([x], [y]) : (x, y) \in Z\}$, and ○ is the operation assigning to each pair $([x], [y])$ of elements of B the element $[x] \circ [y] = [z \oplus w]$, for some $z \sim x$ and $w \sim y$. Notice that $\sim$ in these last two expressions is the relation over X; I shall use the same symbols $\succsim$, $\succ$ and $\sim$ in both structures, since the context will prevent any possibility of confusion. The strategy of the proof of the representation theorem consists of showing that $\langle A, \succsim, B, \circ \rangle$ is an Archimedean, regular, positive, ordered, local semigroup. Lemma 1 below establishes that the relation $\succsim$ and the operation ○ are well defined. The remaining lemmas establish that the structure satisfies the conditions defining local semigroups of the type just mentioned.[11]

LEMMA 1:   *Both $\succsim$ and ○ are well defined.*

Proof: First of all, we want to show that if $[x]$, $[y]$ are elements of B such that $[x] \succsim [y]$, and $x \sim x'$, $y \sim y'$, then $[x'] \succsim [y']$. But this is clear, because the given assumptions imply that $x' \succsim y'$, by (A1).

Next, we want to establish that the result of the operation ○ does not depend on the particular selection of the equivalence class elements, i.e. we want to prove that if $[x]$, $[y]$ are classes such that $([x], [y]) \in B$, and x', y' are elements of X such that $x' \sim x$, $y' \sim y$, then $([x'], [y']) \in B$ and $[x] \circ [y] = [x'] \circ [y']$. The given assumptions imply that there are elements $z, w, z', w' \in X$ such that $z \sim x$, $w \sim y$, $z' \sim x'$, $w' \sim y'$, $z \oplus w$ and $z' \oplus w'$ are defined, $[z \oplus w] = [x] \circ [y]$ and $[z' \oplus w'] = [x'] \circ [y']$. By (A2), we have that $[z \oplus w] = [z' \oplus w']$ and so the desired result follows.   □

LEMMA 2: $\langle A, \succsim \rangle$ *is a simple order.*

Proof: For any $[x]$, $[y] \in A$, either $x \succsim y$ or $y \succsim x$ (A1), which implies that $\succsim$ is connected in A. If $[x] \succsim [y]$ and $[y] \succsim [x]$, then $x \succsim y$ and $y \succsim x$, i.e. $x \sim y$, from which follows that $[x] = [y]$. Finally, if $[x] \succsim [y]$ and $[y] \succsim [z]$, then $x \succsim y$ and $y \succsim z$, which implies that $x \succsim z$ (A1) and so that $[x] \succsim [z]$.   □

LEMMA 3:   *If $([x], [y]) \in B$, and $[x] \succsim [z]$, $[y] \succsim [w]$, then $([z], [w]) \in B$.*

Proof: Since $([x], [y]) \in B$, it follows that there are u, v with $u \sim x$, $v \sim y$ and $u \oplus v$ defined. It follows that $u \succsim z$ and $v \succsim w$, and so there are u', v' with $u' \sim z$, $v' \sim w$ and $u' \oplus v'$ defined (A3). This shows that $(z, w) \in Z$ and thus that $([z], [w]) \in B$. □

LEMMA 4: *If $([z], [x]) \in B$ and $[x] \succsim [y]$, then $([z], [y]) \in B$ and $[z] \circ [x] \succsim [z] \circ [y]$.*

Proof: By Lemma 3, $([z], [y]) \in B$. The desired conclusion follows from (A3). □

LEMMA 5: *If $([x], [z]) \in B$ and $[x] \succsim [y]$, then $([y], [z]) \in B$ and $[x] \circ [z] \succsim [y] \circ [z]$.*

Proof: By Lemma 3, $([y], [z]) \in B$. The desired conclusion follows from (A3). □

For some reason, the proof of one of the most apparently simple properties of a binary operation, associativity, is rather involved.

LEMMA 6: *$([x], [y]) \in B$ and $([x] \circ [y], [z]) \in B$ iff $([y], [z]) \in B$ and $([x], [y] \circ [z]) \in B$; and when both conditions hold, $([x] \circ [y]) \circ [z] = [x] \circ ([y] \circ [z])$.*

Proof: In order to prove the sufficiency part of the biconditional first, assume that $([x], [y]) \in B$ and $([x] \circ [y], [z]) \in B$. Then there are x', y', u, z' such that $x' \sim x$, $y' \sim y$, $[u] = [x] \circ [y]$, $z' \sim z$ and $(x', y') \in Y$, $(u, z') \in Y$. Also there are x'', y'' such that $[u] = [x] \circ [y] = [x'' \oplus y'']$ with $x'' \sim x$ and $y'' \sim y$. It is immediate that $u \sim x'' \oplus y''$ and so (A4) there are x''', y''' such that $x''' \sim x''$, $y''' \sim y''$, $(x''', y''') \in Y$ and $u = x''' \oplus y'''$. Thus, we have both $(x''', y''') \in Y$ and $(x''' \oplus y''', z') \in Y$. Thus (A5), $(y''', z'), (x''', y''' \oplus z') \in Y \subseteq Z$ and therefore, since $[x] = [x''']$, $[y] = [y''']$ and $[y] \circ [z] = [y''' \oplus z']$, we have that $([y], [z]) \in B$ and $([x], [y] \circ [z]) \in B$. The necessity part of the biconditional is proven in an analogous way.

Suppose now that both conditions hold. By (A4), for any $x, y, z \in X$ there are x', y' and z' such that $x' \sim x$, $y' \sim y$, $z' \sim z$ and

$$([x] \circ [y]) \circ [z] = [(x' \oplus y') \oplus z'].$$

Thus, by (A5),

$$([x] \circ [y]) \circ [z] = [x' \oplus (y' \oplus z')]$$
$$= [x] \circ [y' \oplus z']$$
$$= [x] \circ ([y] \circ [z]). \quad \square$$

LEMMA 7: *If* $([x], [y]) \in B$, *then* $[x] \circ [y] \succ [x]$.

Proof: Assume that $([x], [y]) \in B$. Then $(x, y) \in Z$ and so $[x] \circ [y] = [z \oplus w]$ for $z \sim x$, $w \sim y$. By (A6), $z \oplus w \succ z$ and so $[x] \circ [y] = [z \oplus w] \succ [z] = [x]$. $\square$

LEMMA 8: *If* $[x] \succ [y]$, *then there exists* $[z] \in A$ *such that* $([y], [z]) \in B$ *and* $[x] \succsim [y] \circ [z]$.

Proof: Assume that $[x] \succ [y]$. Then $x \succ y$ and so (A7) there are $u, v, z \in X$ with $u \sim x$, $v \sim y$ $(v, z) \in Y$ and $u \succsim v \oplus z$. Hence, $(y, z) \in Z$ and

$$[x] = [u] \succsim [v \oplus z] = [y] \circ [z]. \quad \square$$

For any $[x] \in A$ we define the expression $n[x]$ as follows. If $y_1 \oplus \cdots \oplus y_n$ is defined and $x \sim y_i$ for any i $(1 \le i \le n)$, we let $n[x] = [y_1 \oplus \cdots \oplus y_n]$ and say that $n[x]$ is defined. We also let $N_{[x]}$ be the set $\{n \in N : n[x]$ is defined$\}$.

LEMMA 9: $\{n : n \in N_{[x]}$ and $[y] \succ n[x]\}$ *is a finite set*.

Proof: By definition, for every $n \in N_{[x]}$ there are $y_1, ..., y_n \in X$ with $y_i \sim x$ for every $i = 1, ..., n$, such that

$$n[x] = [y_1 \oplus \cdots \oplus y_n].$$

Thus, to the sequence $1[x], 2[x], ..., n[x], ...$ (which is defined because $m \in N_{[x]}$ for every $m < n$) there corresponds in a one to one fashion a standard sequence $x_1, x_2, ..., x_n, ...$ in X. The condition $[y] \succ n[x]$ implies that this sequence is strictly bounded and therefore (A8) it is finite, which in turn implies that the set $\{n : n \in N_{[x]}$ and $[y] \succ n[x]\}$ is finite. $\square$

Proof of the Representation Theorem: Let $\langle X, \succsim, Y, \oplus \rangle$ be an Aristotelian extensive structure, and define a new structure $\langle A, \succsim, B, \circ \rangle$ as follows. Set $A = X/\sim$ the quotient set of X with respect to $\sim$; for any $[x], [y] \in A$, let $[x] \succsim [y]$ iff $x \succsim y$; let $B = \{([x], [y]) : (x, y) \in Z\}$; finally, if $([x], [y]) \in B$ then $[x] \circ [y] = [z \oplus w]$ for $z \sim x$ and $w \sim y$. By lemmas 1-9, $\langle A, \succsim, B, \circ \rangle$ is an Archimedean, regular, positive, ordered local semigroup. Thus, by Theorem 2.4 in Krantz et. al. (1971), p. 45, there is a function ψ from A to $\mathbf{R}^+$ such that, for all $[x], [y] \in A$,

(i)
$$[x] \succsim [y] \quad \text{iff} \quad \psi([x]) \geq \psi([y])$$

and

(ii) $\quad$ if $\quad ([x], [y]) \in B, \quad$ then $\quad \psi([x] \circ [y]) = \psi([x]) + \psi([y]).$

Moreover, if ψ and ψ' are any two functions from A to $\mathbf{R}^+$ satisfying conditions (i) and (ii), then there exists $\alpha > 0$ such that for any nonmaximal $[x] \in A$,

$$\psi'([x]) = \alpha\psi([x]).$$

Let φ be the function from X to $\mathbf{R}^+$ defined as follows: if $x \in X$, set $\varphi(x) = \psi([x])$. Then we have

$$
\begin{aligned}
x \succsim y \quad &\text{iff} \quad [x] \succsim [y] \\
&\text{iff} \quad \psi([x]) \geq \psi([y]) \\
&\text{iff} \quad \varphi(x) \geq \varphi(y).
\end{aligned}
$$

and also, if $(x, y) \in Y$, then $([x], [y]) \in B$ and

$$
\begin{aligned}
\varphi(x \oplus y) &= \psi([x \oplus y]) \\
&= \psi([x] \circ [y]) \\
&= \psi([x]) + \psi([y]) \\
&= \varphi(x) + \varphi(y).
\end{aligned}
$$

Suppose now that φ' is another function satisfying conditions (i) and (ii) of the theorem. Let ψ' be the function from A into $\mathbf{R}^+$ such that $\psi'([x]) = \varphi'(x)$. Then it is easy to show that ψ' is a representation of $\langle A, \succeq, B, \circ \rangle$ different from ψ. Notice that x is maximal in X iff $[x]$ is maximal in A. Hence, there is a positive α such that for all nonmaximal x

$$\begin{aligned}
\varphi'(x) &= \psi'([x]) \\
&= \alpha\psi([x]) \\
&= \alpha\varphi(x).
\end{aligned}$$

The representation theorem is thus proved. □

The former demonstrations have provided an example of the concept of fundamental measurement, i.e. of the proof of the existence of a numerical structure representing an ontological one. Sometimes —as it turns out to be the case in connection with abstract labor— the representation is not fundamental, because the structure to be represented is already a mathematical one. In these cases, when the structure to be represented is itself representing some real system or things, or is an idealization, the representation is *mediated* by a mathematical structure. This type of mediated representation is very common, e.g., in utility theory, where the existence of utility functions is often established for subsets of the Cartesian space that satisfy certain special conditions. The elements of the corresponding structures are therefore not more or less preferable real bundles of goods, but rather *vectors* representing such bundles. This fact requires the introduction of a concept of representation more general than that of a fundamental one. Obviously, the generalization requires only that the structure to be represented be allowed to be any structure, not just an ontological one.

Chapter 4

THE DIALECTICAL METHOD

The aim of the present chapter is to provide a general philosophical framework for the reconstructive enterprise that has been undertaken in the present book. Usually, those philosophers working in the field of the theory of science, especially in the foundations of some discipline, have had an empiricist upbringing and their way of entering these fields has gone through attempts to solve problems of empiricist philosophy in connection with scientific knowledge. Yet, by no means the concern with foundational problems needs to involve a commitment to empiricist positions. In particular, the framework that I adopt here is not empiricist, and that will have some impact on the formal reconstruction of MTV.

I want to discuss here Hegel's dialectical method and its criticism by Marx, in order to provide my own version of such a method (if it can be called a method at all) within a more general philosophical framework. I will show that the way Marx describes the dialectical method in the *Grundrisse* suggests a procedure to deal with certain aspects of formal scientific axiomatic systems. In the first section I will discuss Hegel's dialectic, whereas in the second I will consider the problem of Marx's "inverted" adoption of Hegel's dialectical method. My own view of dialectic I will present in the final section of this chapter.

4.1 HEGEL'S DIALECTICAL METHOD

What is Hegel's dialectical method? How does it work? In popular literature Hegel's dialectical method is described by means of the triad: thesis-antithesis-sinthesis. As a matter of fact, Hegel never uses this terminology, which is due to Fichte, but in the *Encyclopedia Logic* he asserts that

> in point of form Logical doctrine has three sides: (α) the Abstract side, or that of understanding; (β) the Dialectical, or that of negative reason; (γ) the Speculative, or that of positive reason. [...] These three sides do not make three *parts* of logic, but are stages or 'moments' in every logical entity, that is, of every notion or truth whatever.[1]

This declaration of Hegel's has been taken at its face value by many, and it is widely believed to provide a general description of the dialectical method, not only as it works in the *Logic*, but in Hegel's thought in general as well. Indeed, the famous "thesis" would be none other than the Abstract moment; the "antithesis" would be the Dialectical one; whereas the Speculative moment would be represented by the "sinthesis". Many have been willing to read into the quoted paragraph a rigorous formulation of the dialectical method, and some have even intended to provide a formalization of the same. Most notably, Kosok (1966) has tried to show that the *Logic* can be obtained through a rather rigorous application of a formalized version of the procedure described in the quoted paragraph. Certainly, at least the first triad in the *Logic*, namely Being-Nothing-Becoming, seems to fit the schema in a very precise way, but as one advances in the *Encyclopedia* (even within the *Logic*, which is the first part) the transitions appear to fit less and less such triadic pattern. It is not that such a departure from the original schema should be surprising. After all, Hegel warns in the same section (§79) that "the statement of the diving lines and the characteristic aspects of logic is at this point no more than historical and anticipatory". Furthermore, in the third part of the *Logic*, in the *Zusatz* to §161, it is said that the transition along the three moments is the method applied in the first part, in the doctrine of Being, while other methods, rather different, are applied in the subsequent two parts:

Transition into something else is the dialectical process within the range of Being: reflection (bringing something else into light), in the range of Essence. The movement of the Notion (*Begriff*) is *development:* by which that only is explicit which is already implicitly present.[2]

As a matter of fact, nevertheless, it must be distinguished what Hegel *says about* his method from *how he actually uses* it. As Findlay has pointed out,

The devices by which the Dialectic is made to work are, in fact, inexhaustible in their subtlety and variety. Hegel admits [as we just saw] that they change systematically from one section of the Dialectic to another, but the change is much greater and less systematic than he ever admits. McTaggart, in the brilliant fourth chapter of his *Studies in Hegelian Dialectic*, has gone further in systematizing them than has any other writer on Hegel, and if *he* has failed to reduce them completely to order, it would be vain for anyone else to hope to succeed.[3]

What this means is that the dialectical method (which perhaps should not be called a 'method' at all) is inaccesible to any attempt at formalization. It is impossible to provide a codification of the dialectical transitions (let alone a proof of their "correction") in the style of a formalized logical system (as the one I presented in chapter 3). And the reason for this is deep, because the very aim of the dialectical method is to overcome the rigidity and isolation of the unilateral conceptions of the Understanding. How could any method succeed in this endeavor if it were to use the typical procedures of the Understanding, of which mathematical logic is the epitome? Whereas the application of the axiomatic method requires precise concepts with sharply defined boundaries (idealizations), concepts implicitly and exclusively defined by the axioms and kept apart from other concepts into which "they naturally shade, and without which they can have no significant application",[4] the dialectical method pretends to be precisely a method of concrete reasoning, i.e. a method that unifies the isolated concepts by letting them merge into other concepts with which they are logically related forming *families* of concrete notions. Indeed, if no such families naturally existed, the chains of definitions we find or construct in science would be entirely artificial and would not reflect, not even approximately, corresponding analogous chains of concrete (non-idealized) concepts. In such a case, it would be altogether impossible to connect the concepts thus defined with real phenomena.

This non-formalizable character of the dialectical method is responsible for the reluctance that many philosophers express to accept the existence of something like a dialectical logic. The reasons for this reluctance are compelling, because nowadays the term 'logic' is associated with thought procedures that can be somehow codified and systematized. But if the dialectical "method" is not a logic in the current sense of the term, how can we describe it in general terms?

I shall restrict myself to consider the former question in connection with the dialectical method as it appears in the (*Encyclopedia*) logic, which is after all the doctrine where Marx obtained his own version of dialectic. First of all, it must be noted that Hegel's logic is actually a systematic and thorough presentation of ontology, the kind of discourse that the scholastics put under the heading of '*metaphysica*' or '*prima philosophia*'. In fact, Hegel's logic (ontology) is quite Aristotelian, only that it is presented in a very special way. This way is the path drawn by spirit in its effort to overcome the "contradictoriness" of the concepts it finds in the journey that starts with a consideration of the concept of pure being. This path ends up with the Idea, which includes as "moments" all the concepts of ontology. Hence, in a sense, the Idea is the logical structure of being, it is the complete system of ontology.

Two notions that appear as crucial in the account of dialectic are those of spirit and contradictory concept. Regarding the latter, it is baffling for all those educated in analytical philosophy any talk about contradictory *concepts*, since in the first place 'contradiction' can only be predicated of *sentences* or *propositions*. Yet, if we accept that concepts (predicates) can be defined by means of axiomatic conditions (as they can), then we could define a contradictory concept as one defined in terms of an inconsistent set

of conditions. For instance, we can define the concept of a berkdome in terms of the clause: 'x is a *berkdome* iff x is a dome and
x is round and x is square'. Clearly, the concept of a berkdome
would be then a contradictory one. When Hegel characterizes
his notion of contradictory concept he seems to have in mind
something like this. The most disturbing tenet of Hegel's philosophy is his claim that, moreover, such concepts have instances,
that there are actual contradictions in the world. Yet, again, one
thing is what Hegel *says* about the existence of contradictions in
thought and reality (which sounds like mumbo jumbo) and quite
another the sense of 'contradiction' (*Widerspruch*) as determined
by his actual *use* of the term. According to Findlay,

> it is plain that he cannot be using it in the self-cancelling manner that
> might at first seem plausible. By the presence of "contradictions" in
> thought or reality, Hegel plainly means the presence of opposed,
> antithetical *tendencies*, tendencies which work in contrary directions,
> which each aim at dominating the whole field and worsting their
> opponents, but which each also require these opponents in order to
> be what they are, and to have something to struggle with.[5]

In the sphere of thought, in particular, contradictory concepts
(in the Hegelian technical sense) appear whenever Understanding seeks to give to each of a couple of opposed "reasonable"
concrete concepts (like "what is essential" and "what is accidental") "its distinct empire, or when it sharpens or exaggerates either so as to dominate the whole field and to eliminate its rival".[6]
When a concrete concept is given "its distinct empire" it becomes
isolated, cut-off from those other concepts into which it naturally
merges; when it is sharpened or exaggerated it becomes *distorted*
and one-sided. Idealizations are distorted isolated concepts and
so are contradictory in the already stipulated sense.[7] Indeed, it
is possible to arrest concepts artificially in this way and to stick
indefinitely to them, taking them as the "truth". I shall discuss
below the philosophical conditions of possibility for this to happen, the ontological and epistemological views favored by those
who stick to these concepts as if they were "ultimate truth".

The dialectical method is at work whenever the deeds of Understanding are being emended. Dialectic is the art of following the logical dynamic of concepts, "which determines them to move forward in certain directions when pushed in unwonted ways".[8] This dynamic of concepts is not purely *a priori*, but is grounded in experience, as these concepts also are, since they were grasped in the first place as universal kinds exemplified by objects of experience:

> [...] in the actual working of the Dialectic there is a recourse to experience which is simply a recourse to experience, and which is not based on the demand of abstract argument. [...] In casting about for something that will serve as an opposite, a complement or a reconciling unity of certain phases, Hegel has constant recourse to nature and history: he introduces forms that would never have arrived at through the abstract development of concepts.[9]

As should be plain by now, there cannot be a general formulation of dialectic as a "method" ready to be applied to whatever content turns up. Yet, this does not mean that there is no standard whatsoever in its application. This brings us to the second crucial notion in the account of dialectic, the notion of spirit. According to Findlay,

> the lower categories and forms of being really break down because they are felt to be inadequate approximations to the sort of self-differentiating unity which is to be found only in self-conscious spirit. *This is the secret standard* by which all ideas and performances are judged, and the lubricant without whose secretely applied unction the dialectical wheels and cranks would not turn at all. Anyone who does not feel impelled to think in terms of this sort of self-differentiating unity, will *not* find his inferior categories breaking down, nor leading him to Hegelian results.[10]

Hence, the philosophical pressuposition that makes dialectic work is the operation of spirit in Hegel's sense. Before I provide some criticism of this notion (in §3 of this chapter), it will be necessary to attempt first a characterization of the same.

Hegel's concept of spirit —even if its immediate historical antecedents are to be found in Kant and Fichte— is notoriously similar to Aristotle's concept of *nous*, recovered by the Latin tradition as *intellect agent*. What is new in Hegel is, above all, his endowing of this *nous* of a creative power similar to that of God, except that spirit does not create nature in a *conscious* way (like Jehovah) but develops pursuing its own *telos*, which is self-consciousness. The "essence" of spirit is to strive from "the beginning" toward self-consciousness, and in its endeavor to do so, but only as a blind step necessary to provide itself with an "other", it creates nature, which is thus a lower manifestation of spirit. It is only through man that spirit reaches the character of self-conscious spirit. What this means is that one and the same spirit, intellect agent is present in individual men, which works in them not just as the "universal in action", that mental power that discovers the universal concept in the objects of perception, envisaging these objects as instances of a universal substance-kind, but which also possesses the "absolute negativity of the notion". By this term Hegel understands that capability of spirit in men by which it can abstract from any objective content whatsoever in order to concentrate in itself, to become its own object:

> the essential, but formally essential, feature of mind [spirit] is Liberty: i.e. it is the notion's absolute negativity or self-identity. Considered as this formal aspect, it *may* withdraw itself from everything external and from its own externality, its very existence; it can thus submit to infinite *pain*, the negation of its individual immediacy: in other words, it can keep itself affirmative in this negativity and possess its own identity. All this is possible so long as it is considered in its abstract self-contained universality.[11]

Yet, this is only in its formal aspect, for spirit must continue its journey to become absolute, precisely by following Hegel's *Encyclopedia*, which in this manner constitutes the last stretch to perfect self-consciousness. What this means is that Hegel's logic, the philosophy of nature and the philosophy of spirit are effective parts of this grandiose journey of spirit toward its *telos*; spirit *must* go across Hegelian philosophy in order to reach its end.

In this sense, spirit is the "truth" of everything, since in the end everything is spirit in some degree of development and everything is made by spirit for the sake of its own self-consciousness. Under this light dialectic appears as the work of spirit in the last stages of its journey, as the last part of the effort of spirit toward its finality. Only that this work now produces precisely *concepts*, whereas previously it had produced the world itself. Dialectic is the activity of spirit in producing the conceptual material in order to reach, first, the Idea. In the philosophy of nature spirit acquires consciousness of those inferior aspects of itself previously developed as it moved itself positing nature as a presupposition of conscious mind:

> From our point of view mind has for its presupposition Nature, of which it is the truth, and for that reason its *absolute prius*. In this its truth Nature has vanished, and mind has resulted as the 'Idea' entered on possession of itself.[12]

In the logic spirit obtains consciousness of its own development "previous" to the creation of nature. This is why Hegel says that the logic is "the presentation of God as He was in His eternal essence, before the creation of Nature and finite Spirit".[13] Through the philosophy of spirit, spirit acquires consciousness of itself as *man*, as subjective, objective and absolute mind. What this means is that even man is just a stage, albeit the crowning one, in the way of spirit toward self-consciousness. Individual men are only provisional depositaries of spirit, and their creations —the State, art, religion and philosophy— are at bottom creations of self-developing spirit. It is seen, then, why spirit is "the central notion in terms of which his system may be understood".[14] The apparent obscurity and difficulty of Hegel's philosophy is removed (at least partially) if the concept of spirit as the absolutely fundamental and central notion of his system is kept in mind:

> In terms of this notion many of Hegel's most obscure transitions will become lucid: [their point can be seen] when we realize them to be turns on the path leading up to Spirit.[14]

We shall proceed now to see how Marx pretended to make dialectic work without spirit and without the possibility of any systematic codification of the dialectical transitions.

4.2 THE MARXIAN "INVERSION" OF HEGEL'S DIALECTIC

In the Postface to the second edition of *C*, Marx wrote:

> My dialectical method is, in its foundations, not only different from the Hegelian but exactly opposite to it. For Hegel, the process of thinking, which he even transforms into an independent subject, under the name of 'the Idea', is the creator of the real world, and the real world is only the external appearance of the Idea. With me the reverse is true: the ideal is nothing but the material world reflected in the mind of man, and translated into forms of thought.

Several remarks are in order here. In the first place, Marx is right in saying that for Hegel the process of thinking (*nous*) is an independent subject and creator of the world. Nevertheless, Marx is not careful to point out that the Idea is only a moment of spirit, which is the only true reality. Be that as it may, his claim that the ideal is "nothing but the material world reflected in the mind of man, and translated into forms of thought" never was quite developed or clear. In the first place, what is the "ideal"? Is it the "world" of ideas in the minds of men at a certain stage of history? In the second, what is a "reflection in the mind"? Is it true that whatever 'reflection' means, *every* idea is a reflection of the "material" world? Unfortunately, Marx never provided a further development of these very brief remarks and so it is hard to say what he really meant by them.

One of the crucial notions, the notion of matter, is incredibly obscure, but it seems to have been devised by Marx to expel from his ontology whatever entity that seemed "mystical" to him, quite notably the Hegelian spirit. There does not seem to be any deep philosophical consideration behind this concept of matter, but a rather arbitrary instinct, according to which certain entities are to be allowed as existent whereas others are not. More than a metaphysically developed notion, 'matter' seems to be just a label to attain this purpose. The history of the efforts to develop this embryonic notion of Marx's into a cogent philosophical concept is long and involved. Indeed, Soviet philosophy since the publication of Lenin's *Materialism and Empirio-Criticism*

exhausted every possible definition and argument trying to clarify and defend a Marxist philosophical concept of matter. Unfortunately for Marxism, however, these efforts were prey from the
very beginning to unsolvable inconsistencies and dilemmas that
condemned them to failure. This has been shown in a very detailed way by Lobkowicz (1978), who claims that

> all such inconsistencies and dilemmas are, in the last resort, due
> to the basic paradox of Marxism-Leninism, namely, that it wants to
> be a materialism without leaving the heights of Occidental meta-
> physics which, to Soviet philosophers, is [was in 1963] exemplified
> by Hegel.[15]

Marx's conviction seems to have been that there are only
spatio-temporal entities like bodies, properties and relations
among these entities, which somehow give rise to society and
consciousness. Marx seems to want to leave out of this picture
God, the angels, and any property or relation that is not instantiated by some body or another, or at least causally connected to
the action of some physical or social object. Clearly, such a view
gets rid, in particular, of the Hegelian spirit.

The first step in the inversion of Hegelian dialectic seems to
be thus the claim that all ideas are nothing but reflections of material objects (in the given sense) in the mind of man. The term
'reflection' leaves open a wide room for conceptions about the
way men acquire concepts, but the point seems to be that *all*
these concepts must somehow refer to material objects, on pain
of being meaningless.

The second step in the inversion of Hegelian dialectic seems
to be the recovery of the "general forms of motion" introduced
by Hegel:

> The mystification which the dialectic suffers in Hegel's hands by no
> means prevents him from being the first to present its general forms
> of motion in a comprehensive and conscious manner. With him it is
> standing on its head. It must be inverted, in order to discover the
> rational kernel within the mystical shell.[16]

Whatever the Hegelian "forms of motion" adopted by Marx, it is clear that these forms of motion, which for Hegel were moments of development of spirit toward self-consciousness, are taken by Marx in a rather unsystematic way, deprived of any leading thread. Engels made an effort to present the "laws of dialectic" in the *Anti-Dühring* and the *Dialectics of Nature* but, as Elster (1985) has established, Marx never took seriously these efforts of his partner.[17] As a matter of fact, "[a]lthough he repeatedly intended to set out the rational core of the Hegelian dialectics, he never got around to doing so".[18] Elster finds in Marx (not in Engels) two rather disconnected strands of Hegelian-like reasoning. The first is "the quasi-deductive procedure used in central parts of the *Grundrisse* and in the opening chapters of *Capital I*, inspired above all by Hegel's *Logic*". The second is "a theory of social contradictions, derived largely from the *Phenomenology of Spirit*".[18]

Elster claims that this theory of contradictions "emerges as an important tool for the theory of social change", but I shall not be concerned with it here, since I am more interested in the method of *Capital*, which was declaredly inspired in Hegel's *Logic*. If in Hegel's *Logic* we see the self-development of spirit from abstract being to the Idea, in *Capital* we see the development of the concept of capital from the production process to the concrete price determined by supply and demand. The structural analogies between the first and the second of these processes have been studied by Enrique Dussel in a very careful and detailed way throughout several books, especially in Dussel (1990), which contains the exposition that I shall follow here.

According to Dussel, for Marx the first moment of capital is capital as money, as financial capital ready to be invested in the production of commodities. Thus, notice that here capital plays the role that the absolute played in Hegel's *Logic*, so that if the first definition of the absolute there is 'the absolute is being', in *C* the first definition of capital is 'capital is financial capital'. The negative moment is the negation of capital as financial capital, i.e. 'capital is not financial capital', that is, capital is labor-power and means of production, which is the determined negation of financial capital. The "speculative moment" is the unity of financial capital, labor-power and means of production: this is the productive process, which corresponds to the Hegelian category of becoming. *Daseyn* (*ens*) appears here as commodity, which is the negation of the production process. The first return-into-self is then the negation of commodity, which is now capital as accumulation. In this rather ingenious way Marx develops the schema of his theory of capitalism, proceeding to unfold the content of each of these dialectical moments, in a very detailed way, along *C*.

The former procedure is quite Hegelian indeed, but there is something queer in conceiving accumulation as the identity of financial capital with itself. After all, accumulated capital, even if it is financial capital, has a greater magnitude than the original financial capital with which the cycle had started. Yet, the purely Hegelian method presents this increase of capital as a development of *capital itself*, as something produced by capital alone (remember that the method is just unfolding the "moments" of capital). It is at this point that Marx's departure from Hegel's dialectic appears, even though (as we have seen) Marx makes use of Hegelian procedures. The departure lies, more than in an "inversion" of Hegel's, in a "breaking of the bottom" of his system (that is, of the one that results from a purely Hegelian dialectical development of the concept of capital) by postulating a source (*Quelle*) of value *which lies outside the categories of the system and cannot be sublated by the concept of capital in any way*. This source is none other than live labor (*lebendige Arbeit*), the living worker who is before anything else a person, a human being.

Hence, it turns out that Marx makes use of seemingly Hegelian dialectical procedures just to be able to break the bottom of the resulting product. Strictly speaking, this is the most un-Hegelian way of proceeding, since in this form the source which is postulated cannot be sublated by the totality of the system (the "Idea"), which is to say that somehow it remains isolated. If Dussel is right, Marx produced what he thought would be a Hegelian

political economy just to introduce an element alien to it as creator of value. In the cycle $M - C - M'$ it would seem that the
increase in the returned financial capital M' is a product merely
of capital itself, of the value of the production means and the
labor-power (represented by the paid wage). Marx point is, on
the contrary, that the secret of such increase is to be found in a
factor which lies outside these categories, in live labor.

The systematic expresion of the last two heuristic result is the
Law of Value. It can be said that the whole point of *Capital* is
to prove that, *essentially*, all the different moments of capital are
nothing but live labor, even though they do appear as the self-
development of capital itself:

> Marx tries to show, then, that all the moments of capital (its determi-
> nations: commodity, money, means of production, product, value,
> surplus value, benefit, price, interest, rent, and so on) are, thanks
> to the "Law of Value", objectified "live labor", production of value
> when it is reproduced or replaced; it is creation of value out of the
> nothingness of capital in the case of surplus value.[19]

The whole point of the Law of Value is to show that *every* type
of benefit (in industry, commerce and the land) is nothing but
surplus value, unpaid live labor. Now, since Marx —for ethical
reasons— does not want live labor to be sublated by capital, it is
kept by him out of the system and so the methodological procedure in the development of Marx's theory of capital becomes
instaurated that surplus value —and hence value— must be determined "independently of its form of appearance" (exchange-
value). Since live labor remains isolated from the remaining concepts of the system, it cannot be defined in terms of such concepts, and so Marx is compelled to introduce the concept of live
labor independently of virtually every other notion of capital,
merely as "labor pure and simple, the expenditure of human
labor in general". As we saw along the first two chapters, this
methodological decision lies at the bottom of the foundational
problems of MTV and so we can say that Marx's concept of a
source of value, however plausible it might seem as the philosophical foundation of a humanist critical ethic, as the ground of

the revolutionary overcoming of capitalism, just does not work from a logical point of view. Marx got an ethic but not a satisfactory scientific theory of market economies. We can therefore conclude that Marx's "inversion" or "breaking of the bottom" of Hegel's system is what lies at the basis of his failure to provide an acceptable formulation of the Law of Value. I shall consider in the next section whether a new formulation of the dialectical method can be given, one that avoids the shortcomings both of Hegel's and Marx's dialectics.

4.3 A NEW FORMULATION OF DIALECTIC

4.3.1 THE THEOLOGICAL FRAMEWORK

In spite of the brilliant insights obtained by Hegel in the most diverse areas of philosophy, his characterization of the "process of thought", *nous* or the intellect agent as creator of the world and as unique intellectual principle ("the universal in action") in all men is utterly unacceptable and unsatisfactory. Before any criticism against it is advanced, however, it must be granted that it constitutes the most complete and thorough development of an interpretation of Aristotle's philosophy that arises from a peculiar interpretation of *On the Soul* III-5, where the Philosopher wrote:

> Since in every class of things, as in nature as a whole, we find two factors involved, a matter which is potentially all the particulars included in the class, a cause which is productive in the sense that it makes them all [...], these distinct elements must likewise be found within the soul.
>
> And in fact thought [i.e. the *nous patetikós* or passive intellect], as we have described it, is what it is by virtue of becoming all things, while there is another [i.e. just *nous*, what the Latins later called 'intellect agent'] which is what it is by virtue *of making all things:* this is a sort of positive state like light; for in a sense light makes potential colours into actual colours.
>
> Thought in this sense of it is separable, impassible, unmixed, since it is in its essential nature activity (for always the active is superior to the passive factor, the originating force to the matter).

Actual knowledge is identical with its object: in the individual, potential knowledge is in time prior to actual knowledge, but *absolutely
it is not prior even in time*. It does not sometimes think and sometimes
not think. When separated it is alone just what it is, and this above
is immortal and eternal [. . .], and *without this nothing thinks*.[20]

This passage has always been a matter of much debate but it
is clear that at least under some interpretation it describes *nous*
in terms which are similar to Hegel's characterization of spirit:
nous is active, maker of all things, identical with its object, prior
to potential knowledge in men, always thinks (its nature is thinking), it is immortal and eternal, and no man can think without
it. It is pretty clear to me that Hegel's system can be profitably
seen as the most serious attempt to rebuild the whole of Aristotle's metaphysics upon this conception of *nous*. Yet, even though
Hegel was the first to stress the *creative* aspect of *nous*, putting
its self-development as the backbone of his system, certainly he
was not the first one in claiming that *nous* was one and the same
intellectual principle in all men. In point of historical fact, Averroes was in the past the champion of this view and, indeed, most
of the motivation of St Thomas Aquinas' effort to reconstruct
Aristotle's philosophy sprang from the need felt by the Christian philosophers to refute Averroes. Averroes claimed that *both*
the passive and the active intellects were respectively unique and
separated from the human soul, so that there is only one passive
intellect for all men and only one active intellect for them all:
my intellects are numerically identical to yours and to those of
any other man. In *De unitate intellectus contra Averroistas* Aquinas
makes clear that Averroism is repugnant to Christian faith (*re-
pugnet veritati fidei christianae*), asserting that

once the diversity of the intellect is subtracted from us, which alone
among the parts of the soul appears as immortal, it follows that after
death nothing of the soul of men remains, except the unity of the
intellect; and thus the retribution of prizes and punishments, and
their diversity, would be supressed.[21]

In this work Aquinas attempted to prove that the Averroist
position is "at least as contrary" to the principles of (Aristotle's)

philosophy "as it is against faith".[21] Yet, his target in this opus-
culum is not Averroes' claim that the intellect agent or active
is separated, as his claim that the passive one is so. His arguments against the separateness of the active intellect he gives in
the *Summa Theologica* (Ia, 79, 4-5) as well as in the *Summa contra
Gentiles* (2, 76). Aquinas interprets the *De anima* passage quoted
above as teaching the individual character of the intellect agent
in individual men, and provides arguments to sustain the truth
of such a teaching. His main argument, which must also be directed against Hegel's notion of the "universal in action" is that
if *nous* were the same in all men, its functioning would be independent of the will and control of the individual and —since
its very essence is thinking— it would be continuously thinking;
but it is clear by experience that we can pursue or abandon our
intellectual activity at will. Another argument is that a separated
active intellect would be more perfect that one limited by sensa-
tion, and so it becomes hard to understand why such a separated
agent would need the help of the senses to perform its function
(which, as the interpretation of Aristotle's passage claims, per-
forms anyway previous to the thinking of individuals). Moreover, if it can perform it that way, then, being numerically the
same in men or separated from them, how could it get the limi-
tation of requiring the senses of men to grasp the universal? This
difficulties constitute —in my view— very strong reasons to reject that interpretation (even if it is what Aristotle really meant
to say) and stick to Aquinas' view that the active intellect is not
universal, but each man has his/her own individual active (and
also passive) intellect as constituting an essential aspect of his/her
being.

Hegel's way out of the former objections (which, it seems to
me, does not answer the one advanced by Aquinas) would be
to grant that, indeed, that germ of spirit that self-develops into
the higher forms of consciousness *cannot act as nous except through
the senses of men*, that it requires to produce such senses in order to become —properly speaking— the universal in action. In
other words, men are necessary instruments of the Idea in order

to get consciousness and self-consciousness. As Westphal (1989) has shown, Hegel proves in the *Phenomenology of Spirit* that self-consciousness is possible iff individual human beings are conscious of objects.[22] This means that there is no other form for the Idea to become self-conscious than through experience of the world posited by itself as precondition for this to happen.

As can be seen, the prodigious coherence of Hegel's system makes it hard to find cracks through which such a disturbing idea of a being that produces its own determinations in order to think itself can be criticized. As a matter of fact, the Marxist attempt to crack Hegel's system by means of that little poor and confused notion of matter is pitiful. Appeal to the "intuition" that there cannot be such a thing as the self-developing Idea is not at all an argument against the solidity of Hegel's system, but rather a renouncement to philosophy, unless we restrict philosophy to those petty analyses in which empiricism takes so much pride.

For those who believe that there is something valuable in the basic view of the world afforded by Aristotle, the confrontation with Hegel's system is unavoidable, as Hegel is one of the most coherent Aristotelians that have ever existed. Indeed, *basically* the only other interpretation is the one provided by (roughly) Aquinas and the scholastics. And this is no accident. As Schelling noticed in his time quite clearly, the only way to notch the knife of Hegel's spirit is to oppose to it the sword of the Word of God as *revealed* in the Scriptures. In his *Toward a History of Modern Philosophy*, Schelling synthesized in the following form the global vision of Hegel's philosophy:

> God, the Father, before all, is the pure logical concept, which identifies itself with the pure category of *being*. That God must manifest itself, because his essence includes that necessary process; such revelation or alienation of himself in the world is God, the Son. But God must sublate or bring back on himself that alienation: it is the negation of his pure logical being: negation that is accomplished through humanity in art, religion and fulfilled in philosophy; that human spirit is equally the Holy Spirit, by which God takes for the first time consciousness of himself.[23]

Quite against Hegel's pantheism of a *nous* that thinks itself, Schelling claims in his *Philosophie der Offenbarung* that the Creator is previous to being, and is beyond, being an other-worldly reality. Perhaps to prevent the revealed concept of God from getting confused with the absolute defined by the motion of the Hegelian categories, Schelling says that God is not being, but rather the creator of being and his Lord, the *Herrn des Seins*.[24] This view that the creator of being is not itself (or Himself) being appeared for the first time in Plotinus.[25] It was also for the first time refuted by St Augustine, who asserted that "the creator of being is".[26] Moreover, according to the same Revelation that Schelling is appealing to, the very *name* of God is 'being':

> [...] God said to Moses, "I AM WHO I AM" [...] This *is* My name forever, and this *is* My memorial to all generations.[27]

Schelling seems to have thought that that move was required (even though in fact it is not) in order to avoid —among other things— the unacceptable consequence that the human spirit is "equally the Holy Spirit". Most certainly, according to Revelation (and experience), the human spirit is *not* the Holy Spirit. Moreover, contrary to what Hegel's system seems to imply, namely that "God" (the Absolute Spirit) is a sort of rational result which speculative philosophers can reach by following Hegel's system up to its consummation, the Scriptures teach that even the least intellectual men can be saved, because salvation has nothing to do with *merits*, be they intellectual or otherwise, but with *faith*. According to Revelation, it is only by accepting Christ Jesus as Lord and Savior that men can be saved. And only those who are redeemed in this way can be regenerated in their spirit, in particular in their intellect, as a necessary condition to understand heavenly matters. Hegel seems to suggest that heavenly matters can be understood by speculative methods alone, thus making of faith an unnecessary accessory.

Kaufmann claimed that Hegel

> tried to do from a Protestant point of view what Aquinas had attempted six hundred years earlier: he sought to fashion a synthesis of Greek philosophy and Christianity, making full use of the labors of his predecessors.[28]

Unfortunately, if that is what Hegel was trying to do, he failed, as can easily be seen from the Scriptures. His claim that revealed religion and speculative philosophy have the same content, except that the first uses "pictorial language" is just not correct. Speculative philosophy cannot be taken as a good elucidation of the Bible. Another consequence of Hegel's notion of spirit is the suggestion that salvation is a political endeavor, since it seems to consist merely in the emergence of spirit in the form of a "Christian" community and an ethical State, something which seems to depend only on the will of that supraindividual entity. This clearly goes against the Christian teaching according to which salvation is a purely individual endeavor, and has as result eternal life for the *individual* that has been saved, even though the redeemed constitute a godly people in the form of visible communities. Very much in an Averroist fashion, Hegel's philosophy does not make room for eternal life for the individual, but only for spirit. No matter what Hegel might have said on the contrary, what else beside this supraindividual spirit could remain after the death of the individual men?

Avineri (1972) clearly shows how Hegel ascribed to the political situation of the decadent Roman Empire the origin of the category of individual as required by Christianity to prosper as a massive religion. According to Hegel, Christianity could only thrive in a historical situation in which men were deprived of their political rights by the Emperor, being reduced to mere holders of private property. In contradistinction to the ancient *polis*, in which the identity of the citizens was the State and so they were ready to die for it, in decadent Rome the citizen was able to identify himself only with his own property —a very finite and transient thing— and that is how the fear of death arose.[29] This is very suggestive of the role Hegel attributes to the ethical State, which is something like a very earthly realization of the Kingdom of God.[30]

The idea that the Kingdom of God has to be realized historically on earth like a political institution seems to have inspired Marxist communism. Once deprived of the concept of *Geist* and inverted, what remains of Hegelian philosophy (Marxism) can do without the "pictorial language" of religion and go all the way to claim that it is necessary to realize on earth something like an analogue of the Hegelian heavenly kingdom, something like the Kingdom of God, but inverted. If Hegel's philosophy of mind reduces the worth of the individual to that of being a (provisional) bearer of spirit, in Marxist philosophy the worth of the individual is reduced to nothing, since every allusion to "mystical" entities and other-worldly realities is out of the question. This is the very essence of Totalitarianism, because a logical consequence of this inverted Hegelian view is that there cannot be anything more valuable than "the totality", i.e. a purely human State, a State that is not accountable to anything else, not even to Absolute Spirit!

Against this despicable view of man, and unless we are prepared to step again on the slippery slope to Gulag and Buchenwald, it is necessary to assert the integral individuality of the human soul, including the intellect, and the accountability of all men (and the State) to a transcendent Holy God (the God, I believe, of Israel and Christ Jesus). This is the only absolute foundation for the infinite worth of each individual man or woman. This is also the absolute foundation of religious freedom, because the conception of the human spirit that it involves makes room for all faiths, as well as for those who have not had any religious experience by which they should at all come to believe. In my view, this is the theological presupposition for a fresh reconsideration of the insights and novelties introduced by Hegelian dialectic.

4.3.2 DIALECTIC REVISITED

The central notion of Hegelian dialectic, the "lubricant without whose secretely applied unction the dialectical wheels and cranks would not turn at all", the concept of *Geist*, is not acceptable on theological and metaphysical grounds. Yet, even though no "interpretation" of Hegel is acceptable that does away with

Geist, there is no doubt that Hegel has been one of the greatest philosophers ever and that he has very important results and illuminating insights. How can we make use of these results without falling back into *Geist*? In a very strict sense, this is impossible, because these results cannot be understood properly out of the Hegelian system, in which they belong. The most we can do is use them as suggestions to build *other* philosophical theories, without pretending that they are Hegelian except in the wide sense in which Haydn or the young Beethoven can be said to be Mozartian. Certainly, Hegel's theory of the State,[31] his account of civil society,[32] his ethics,[33] his critique of Kant's conception of the object coupled with his own view,[34] his overcoming of skepticism through a brilliantly argued epistemological realism,[35] are all magnificent philosophical masterpieces even though *Geist* is quite an indigestible item.

The method that I propose to recover whatever can be recovered from Hegel's valuable theories is to substitute 'human active intellect' for *Geist* whenever the context indicates that Hegel is speaking of the human mind, and see what turns up.[36] It is obvious that many things will have to go but also many insights can result. It will be interesting, for instance, to see what turns up if this procedure is followed in Kenneth R. Westphal's reconstruction of Hegel's argument for epistemological realism in the *Phenomenology of Spirit*. Even though this argument is essential to show the inadequacy of empiricism as a theory of knowledge (according to Hegel, empirical knowledge or "sensuous consciousness" is the roughest and most elementary form of knowledge), I shall not be concerned with it here so much as with Hegel's theory of universals as substance-kinds.

According to Kant, the objects that we find in (sense) experience are unities constructed in the transcendental subject, in applying the categories to the manifold of intuitions. Hence, it does not make sense to talk about objects as they are "in themselves", since the unity and structure objects have is nothing but the result of the synthesizing activity of the understanding. According to Kant, the categories applied by the understanding are *a priori* and definite in number. But there is only one step from here to consider *all* concepts as playing a role in the synthesizing activity of the transcendental subject. This leads to a conceptual relativism that makes dependent the nature of objects on the peculiar conceptual apparatus that we have available at a certain moment. Clearly, since these objects are somehow a creation of the concepts we happen to have, the finer the concepts the sharper will be our objects of experience. It is clear that to a theory of knowledge of this kind the problem of the rigidity of the conceptions that Hegel called "of the understanding" just does not arise. Those who maintain this view are thus able to stick indefinitely to determinations which otherwise are clearly abstract and idealized.

The demand for extreme standards of rigour in science and philosophy, usually coupled with deep contempt against any notion or procedure not amenable to logico-mathematical treatment, and a thoroughgoing (if sometimes clandestine) attachment to this Kantian view of the object, or to an empiricist one, springs from a reluctance to deal with the difficulties of dialectical thinking, to accept the possibility of a mode of thought not controlable by mechanical or standard devices. Indeed, some paroxistic versions of this view even postulate as a main task of philosophy the "clarification" of concepts, meaning by this that philosophy should increasingly get rid of all those "vague" and "imprecise" concepts, replacing them by their corresponding "elucidations", which are (and can be) nothing but abstract idealizations of concrete, albeit harder to handle notions.

Since barely only (if any) the inert or relatively simple objects usually studied by physical science can be adequately thought by means of precise concepts, or they can be thus conceptualized without too much distortion, usually the philosophers of science associated to this view tend to concentrate almost exclusively in physics, and, since the objects and concepts of the social sciences tend to be inerradicably vague, there is a tendency to neglect the treatment of these sciences or to dismiss them as utterly "unscientific".

The so-called "problem of incommensurability", for one, is an (undesirable) consequence of this view of the unity of the object. This problem arises for transcendental idealism because it cannot see the objects but as creations of the intellect through given concepts. Thus, objects constructed with different concepts must be different objects as well. In this form, transcendental idealist philosophers of science become baffled by the outcome that special relativity and classical mechanics have no objects in common! But everybody else knows that many of the phenomena (strictly the same) treated by one theory are also treated by the other, even though there may be some differences in the way they are described.

The problem of incommensurability is a Kantian problem, because it arises from the attempt to represent the scientific phenomena in terms of purely idealized concepts, pretending that it is impossible to refer to them in a way independent of the abstract theory in question, namely in terms of the substance-universals they exemplify (if they have a real existence at all). The problem of incommensurability disappears once we reject the Kantian view as untenable and realize that not all concepts have the same status. This is one of the central tenets of Hegel's epistemological realism: both the rejection of Kant's conception of the object and his claim that the unity individual objects have does not depend upon the unity of apperception (of the subject), but is due to the very nature of these objects, which exist as instances of indivisible substance-universals:

> [. . .] according to Hegel, objects are not in fact mere 'combinations' of sensible properties, as the Kantian model suggests, and on which his doctrine of synthesis depends. Instead, as we shall see, Hegel argues that individual objects exist as manifestations of indivisible substance universals, which cannot be reduced to a set of properties or attributes; he therefore holds that the object should be treated as an ontologically primary whole. As a result Hegel adopts a metaphysical picture which enables him to argue that the object forms an intrinsically unified individual: because the individual is of such and such a kind (a man, a dog, a canary) it cannot be reduced to plurality of more basic property-universals, while it is the universal

that confers this substantiality upon it. In this way, Hegel replaces Kant's 'bundle' model of the object with a more holistic picture, which treats the individual as a unity, in so far as it exemplifies a substance-kind. It is this ontology of substance which lies behind his rejection of the latter's doctrine of synthesis.[37]

It is in this sense that substance-universals can be said to be "structures in the world", and so whenever one grasps one of these universals one is grasping the nature of any of its instances. I claim that these substance-universals are the less idealized concepts available, even though some of their instances are "truer" than others. The status of substance-universals is quite different from the status of idealized concepts. The concept of *homo oeconomicus*, for instance, is not a substance-universal and, in fact, it does not even have instances. It is a fictitious entity that behaves all of the time and exclusively in a form in which human beings barely do sometimes under very special circumstances.

Even though I do not believe that there is an entity that posits itself as being and then begins to engender the categories of Hegel's logic (ontology), I still believe that that ontology is crucially important because it provides the foundation for the doctrine of the notion, in which a non-Kantian theory of essential predication and the object emerges, as Stern (1990) has clearly shown. I propose to see the logic —and dialectic in general— as the development of categories that break down from the point of view of the human intellect just because they are inadequate in view of unwonted experiences and/or an implicit undeveloped view of a totality. From this point of view the "Idea" is not God but only a conceptual construction that constitutes the system of ontology, the "logical structure of the world in its relation to God". What I suggest is that what really moves the dialectic is the desire intrinsic to the human intellect to reach systematic and concrete totalities of thought, to put isolated notions in a wider unifying context, to relate such notions —and the totality as well— to factual reality. Hegel's dialectic is appealing and immortal as a *cantata* of Bach's because it constitutes one of the greatest efforts to fully satisfy this drive. This very drive of the intellect toward concreteness and system is witnessed in action in the tendency to get rid of idealizing assumptions in economic theory. We shall see and explain in some detail this dialectic in the final subsection of this chapter.

4.3.3 THE DIALECTIC OF THE THEORY OF VALUE

As we saw in §2, Marx never got around to set out the "rational core" of the Hegelian dialectic. Marx was always very brief in his written declarations about the nature of the dialectical method. As it was correctly pointed out by the late Professor Jean van Heijenoort,

> Marx has not bequeathed us explicit teachings on dialectic compa-
> rable in extent and precision to his economic doctrines. On dialectic
> the great theoretician left, on the one hand, fragmentary formula-
> tions scattered in his works and correspondence and, on the other
> hand, the product of his dialectical method, the monumental *Cap-
> ital*. We have the fruit of the method, but no systematic exposition
> of the method itself.[38]

There is no doubt whatsoever that Marx thought of doing such an exposition. For instance, in a letter addressed to Engels, dated January 14, 1858, Marx wrote the following:

> If there should ever be time for such work again, I should greatly
> like to make accesible to the ordinary human intelligence, in two
> or three printer's sheets, what is *rational* in the method that Hegel
> discovered but at the same time cloaked in mysticism.

"Unfortunately —adds Van Heijenoort— Marx died without having written those two or three printer's sheets which doubt-less would have forestalled many subsequent discussions".[39] This does not mean, however, that Marx did not write anything at all about dialectic. The work where he explained himself most in this respect was the *Foundations of the Critique of Political Econ-omy*, also known as the *Grundrisse*, written in 1857. Section 3 of the *Introduction*, labeled "The Method of Political Economy" is particularly important in this connection, since it is there where Marx characterizes the dialectical method in the most explicit way, namely as a method that consists in "rising from the abstract to the concrete". Regarding this paragraph, Van Heijenoort said the following:

> [the same] represents in my opinion the most important method-ological document we possess to fill the void left by the absence of those "two or three printer's sheets" on dialectic which Marx never had the leisure to write.[40]

This document represents the most explicit statement of the quasi-deductive deductive procedure used in the *Grundrisse* itself and the opening chapters of *C* but, as we saw in §2, Elster claims that there is yet a second strand of Hegelian dialectic in Marx's thought, namely a theory of social contradictions largely derived from the *Phenomenology of Spirit*. Thus, at the very least it is clear that such document is the most explicit presentation of one of the strands of Hegelian dialectic in Marx's thought, and hence it is possible to argue that *what legitimatelly must be understood by that strand of Marx's dialectic is precisely the contents of such document.*

As Hamminga (1990) has shown in a very detailed way, the methodology that Marx pretended to apply in *Capital* can be naturally seen as a dual motion that builds idealized models and then proceeds to eliminate the restrictive assumptions defining them, in order to yield less idealized models, maintaining at the same time the validity of the original fundamental laws even within these less idealized models. We shall also see that this dual motion can be analyzed in terms of the concept of rising or pass-ing to the concrete.

I devote the first part of this subsection to propose a new in-terpretation of the process of "rising to the concrete", in logico-mathematical terms, trying to stick as far as possible to the text of "The Method of Political Economy"; to that effect, I use an easy example taken from classical physics. In the second part I take as point of departure certain results in Nowak (1980), and the history of Marx's MTV, to propose an elucidation of the afore-mentioned dual motion as well as to interpret the situation of this theory previous to the developments introduced in the sub-sequent chapters of this book. In the third part I discuss the dialectical method again, in connection with this particular interpretation of it. I close this chapter arguing that Marx in fact failed to apply correctly this method, due precisely to his ethical reluctance to sublate the concept of living labor in the category of capital, which makes that concept into an abstract, isolated one.

4.3.3.1 DIALECTIC AS MODEL CONSTRUCTION

In the already mentioned "The Method of Political Economy"[41] Marx characterized the dialectical method —in a very Hegelian fashion, as we can see— as the one that, departing from abstract determinations, "rises to the concrete" (*vom Abstrakten zum Konkreten aufzusteigen*) reproducing the real concrete in the process of thought (*im Weg des Denkens*) as a concentration of multiple determinations, as unity of the diverse (*Zussamenfassung vieler Bestimmungen, Einheit des Mannigfaltigen*). In the sections previous to the one mentioned, Marx applied this method —which he considers as "the correct scientific method"— to elaborate an economic discourse in which the concepts of production, distribution, exchange and consumption are presented as "the articulations of a totality, differentiations within a unity".[42] This discourse is interesting but its incipient character does not make it apt to illustrate the application of the dialectical method that I want to present in this section. According to this application, the procedure of passage from the abstract to the concrete consists, fundamentally, in the construction of singular models of given scientific theories in order to represent determined real concrete situations. I will illustrate what I mean in what follows by means of a relatively simple example taken from classical mechanics.

As a case of "reproduction of the concrete in the process of thought" (*Reproduktion des Konkreten im Weg des Denkens*), consider the problem of constructing a physico-mahematical representation of the phenomenon of free fall of a steel sphere which is left to fall from the roof of a building. In this case the real concrete can be identified with the motion of the body since the moment in which it starts falling until it hits the sidewalk for the first time. The "abstract determinations" (*abstrakten Bestimmungen*) which serve as point of departure are the following.

(I) A set P having as a unique element the center of the steel sphere, which is said to be a material point or particle, i.e. it is assumed that the mass of the body is concentrated in its center.

(II) The concept of time of motion of the particle, which is identified with a closed interval of real numbers $T = [0, t^*]$, where t^* is a parameter that measures the number of seconds that the motion lasts.

(III) The concept of instantaneous position, which is a function $\mathbf{r} : P \times T \rightarrow \mathbb{R}^3$ assigning to the particle and each instant $t \in T$ a vector in the linear space $\mathbb{R}^3$. It is usual and convenient in this case to identify the unidimensional subspace of $\mathbb{R}^3$, generated by the unit vector $\mathbf{k} = \langle 0, 0, 1 \rangle$ (the z-axis), with the straight line determined by the point where the falling starts together with the point on the sidewalk hit by the sphere. It is also convenient to assume that this last point is the origin of the space determined by the straight line on which the falling takes place and the plane through the origin orthogonal to this line.

(IV) The concept of mass of the particle, which is represented by a positive real number m.

(V) The concept of acceleration due to terrestrial gravity, which is identified with the constant number g, and whose value depends on the point of the Earth where it is measured, but which in the present case we can take to be equal to 980 cm/s^2.

(VI) The concept of forces exerted on the particle, which are identified with vectors in the linear space $\mathbb{R}^3$. Apart from a mutiplicity of forces whose resultant is small and negligible, it is usually assumed that there are two forces acting on a freely falling body, to wit, the force of gravity and the resistance of the air. The available information about the phenomena of free fall suggests, however, that in this particular case it is even possible to neglect the air resistance, by which the number of forces to consider is reduced to one, which is none other than the weight of the sphere, and hence mg.

(VII) The definition of the concept of acceleration as the second derivative of the position function with respect to time, i.e.

$$\mathbf{a} = \frac{d^2\mathbf{r}}{dt^2}.$$

(VIII) Newton's Second Law, which asserts that force is equal to mass multiplied by acceleration:

$$\mathbf{F} = m\mathbf{a}.$$

(IX) Concepts and techniques pertaining to differential and integral calculus, as well as to linear algebra.

Out of (I)-(IX) equation

(1)
$$m\frac{d^2\mathbf{r}}{dt^2} = -mg\,\mathbf{k}.$$

is obtained and thereafter, by means of (IX), the function $\mathbf{r}$ as an explicit function of t is obtained by integrating out the right hand side member of (1) divided by m:

(2)
$$\mathbf{r}(t) = (-\frac{1}{2}gt^2 - v_0 t + r_0)\mathbf{k}.$$

Expression (2) is still undetermined, since the specific values of v_0 and r_0, which represent respectively the initial speed and the distance from the sidewalk to the roof, have not been established yet. These values are not abstract determinations, but rather measurements pertaining to a concrete situation. Nevertheless, they are essential to obtain the numerical determined representation of the phenomenon subject to study. What this means is that beside the "abstract determinations" empirical data about the situation subject to study are necessary in the construction of singular models. An examination of this situation and appropriate measurements reveal that the initial speed v_0 is zero, whereas the distance r_0 from the place where the body begins to fall to the sidewalk is (say) 10 m. Using these data the following function is obtained in perfectly explicit determined and concrete terms:

(3) $$\mathbf{r}(t) = (-\frac{1}{2}gt^2 + 1000)\mathbf{k}.$$

Adding one more empirical datum, the value of the mass of the body (let us say one kilopond), we get the relational structure

$$\mathfrak{P} = \langle P, T, \mathbf{r}, m, \mathbf{f}, g \rangle,$$

where P is the singleton whose unique element is the sphere; T is the interval $[0, (2000/g)^{1/2}]$;[43] $\mathbf{r}$ is the vectorial function given by (3); $m = 1$ kp; $\mathbf{f}$ ($=\mathbf{F}$) is the vector $(-980 \cdot 10^3)\mathbf{k}$, which represents a force of 980 thousand dynes directed toward the sidewalk; and g is the adopted value of terrestrial gravity. The structure $\mathfrak{P}$ is a physico-mathematical representation of the situation we have been considering; it is a "totality of thought" (*Gedankentotalität*) which, in fact, reproduces the real situation as a "concrete of thought" (*Gedankenkonkretum*). Departing from abstract determinations and empirical data, we have in effect reached the reproduction of the concrete real —the falling of the sphere from the roof— in the process of thought, that is to say, of empirical and mathematical reasoning, as a synthesis of multiple determinations, as unity of the diverse. This synthesis or concentration of multiple determinations, is none other than the relational structure $\mathfrak{P}$, whose unity is the unity corresponding to an object, pertaining to the universe of set theory, that satisfies certain special conditions (laws); i.e. it is a structure in the sense defined in chapter 3.

In general, in this specific application, the dialectical method of passage to the concrete consists in constructing singular models of real situations out of abstract concepts and empirical data, by means of laws pertaining to a determined scientific theory, which is assumed as something previously given. In the example given the model represents a real situation but in some cases —as we shall see in the next section— it is useful or necessary to construct very idealized models not having a real counterpart, except in the sense that they are a quite remote approximation to a real situation.

Before passing to the next section, perhaps it will be convenient to say a few words about the sense that the terms 'abstract' and 'concrete' have received in the present context. Sometimes the term 'abstract' is taken as a second intention term to refer to objects such as concepts, propositions, sets, or numbers, i.e. what some authors call 'abstract entities'. According to this use of the term 'abstract', the term 'concrete' is used to designate just that which is not abstract, i.e. real beings. It must be clear that Marx's usage of these terms does not follow those conventions. According to the interpretation offered here, abstract determinations are actually for Marx abstract entities (in the current sense nowadays), since they are objects such as propositions and concepts. Nevertheless, Marx distinguishes the "concrete real" from the "concrete thought": The concrete real is a real being (and hence concrete in the current sense nowadays), whereas the concrete thought is a conceptual structure (and thus abstract in the current sense). For Marx —as for Hegel— the distinction between the abstract and the concrete, in the plane of thought, is a distinction between degrees of complexity and articulation. For instance, the concept of mass is abstract with respect to the models of classical mechanics, because it is a concept that by itself does not provide a complete and finished comprehension of any particular reality, i.e. it is being taken in an isolated way, without its being specifically articulated in any of the totalities-of-thought that correspond to it (the models of the theory to which it belongs).

4.3.3.2 DIALECTIC AS THEORY CONSTRUCTION

According to Leszek Nowak, the Marxian method of passage to the concrete coincides with what he calls "method of idealization":

> The method which is able to reveal inner connections is called by Marx the method of proceeding from the abstract to the concrete, i.e. it is the method of idealization according to my interpretation.[44]

According to Nowak the method of idealization consists in the postulation of a series of idealized (false) assumptions about the object under study, and in showing that within these assumptions a particular case of a scientific law is satisfied. As an example of an application of this method, Nowak proposes the formulation by Marx of the Law of Value in *Capital* as an "idealizational sentence", i.e. as a counterfactual conditional of the form:

$$(T^k) \quad \text{If} \quad G(x) \wedge p_1(x) = 0 \wedge \cdots \wedge p_{k-1}(x) = 0 \wedge p_k(x) = 0$$
$$\text{then} \quad F(x) = f_k(H_1(x), ..., H_n(x)).$$

"where $G(x)$ is a realistic assumption while $p_1(x) = 0, ..., p_k(x) = 0$ are idealizing assumptions $(k > 0)$".[45] In the case of the theory of value Nowak says that the propositional function $G(x)$ defines the "universe of discourse" of the theory, whereas $p_1 - p_k$ (for $k = 8$) are simplifying conditions that define an ideal economic system, similar to the "ideal gases, perfectly rigid bodies, and other constructs of the type".[46] According to Nowak, Marx's method in *Capital* consisted precisely in the construction of a sequence of conditionals $T^8, T^7, ... AT^2$, that departing from T^8 eliminated the idealizing assumptions until it obtained a conditional T^2 in which only two simplifying assumptions remain and whose consequent is a modification (in fact a generalization) of the original version of the Law of Value. The conditional AT^2 asserts that if the antecedent is satisfied then the Law of Value holds *with a certain degree of approximation*, so that AT^2 is a weakening of T^2. This procedure is called by Nowak 'concretization'.[47]

We shall see that the procedure of concretization is a particular cases of passage to the concrete, but at this point it is important to prevent the terminology from confusing us. Irrespective of the terminology that is deemed as more adequate (I shall discuss this point later), it seems to me that the procedure Marx attempted to follow in *Capital*, as well as the recent history of Marx's theory of value, can be adequately described as a process quite similar to what Nowak calls 'concretization'. In terms of models and structures, what Marx did in *Capital* was to describe a series of more or less idealized models and to presume that the so-called Law of Value (never formulated precisely) should be satisfied in such structures. I say that Marx *presumed* such a thing because, certainly, he never *proved* that the mentioned law was satisfied in *any* of those models (in fact it can be proven to hold under certain conditions, as we saw in chapter 1, in the prototype of MTV). But not even in the prototype, which is the most idealized of them, would have been Marx able to prove the Law of Value, since the explicitation of the assumptions required for such endeavor required a development of the conceptual apparatus of mathematical economics that did not take place until the middle of the current century, thanks to the work of Samuelson, Leontief, Arrow and Koopmans.[48] As we saw in chapter 2, making use of such conceptual apparatus Michio Morishima (1973) provided the first complete listing of the assumptions required in the derivation of the Law of Value —thus defining precisely the linear model of the theory of value, also known as the Leontief model or the prototype of MTV— and effectively derived the Law of Value from these assumptions, in a way similar to the one we followed in chapter 1. Taking as point of departure this model, Morishima developed the totality of Marx's economic theory in a rigorously scientific way, although of course within the limitations imposed by the very idealized asssumptions upon which the proofs concerning the existence of unique positive values are based. As we saw in chapter 2, these assumptions (most of which were taken for granted by Marx himself in *Capital*), were

severely criticized by Morishima in the same book, whose conclusion is that Marx's theory of value (i.e. what really is the prototype of the same theory) should be abandoned and replaced
by a theory combining aspects of (the linear model of) Marx's
theory and of that of Von Neumann.

   In the terms already discussed in the second chapter, Morishima himself (1974) addressed the problem of constructing a
model of a "new" theory that reinterprets Marx in Von Neumannian terms. Unlike the classical model, this model allows a
better treatment of the problems related to the age structure of
capital goods when the time factor is introduced (which seem
to be unsolvable within the classical model), and admits joint
production as well as choice of techniques. Morishima was able
to prove both the Law of Value and the Fundamental Marxian
Theorem within this Von Neumannian model using "optimal"
instead of "real" values. If real values are obtained in the linear
model computing the contents of labor incorporated into the
commodities on the basis of the prevailing technical coefficients,
optimal values are shadow-prices determined by a linear program which is dual to another linear program for the efficient
utilization of labor. Even though optimal values are not necessarily unique, the exploitation rate is well defined and —as I
pointed out before— the Fundamental Marxian Theorem can
be proven, under the assumption that labor is homogeneous.
We also saw in chapter 2 that, along the same lines, John E. Roemer (1980, 1981) produced a series of more general models and
derived the existence of Marxian equilibria from the assumption that the behavior of the firms consists of maximizing profits
given a set of possible processes of production and certain restrictions in the availability of capitals. In Roemer's models the
value of a bundle of commodities is defined as the minimum
labor required to produce the bundle, given the technological
possibilities of the economy. It is thus seen that Roemer's definition of value is analogous to that of Morishima's, the difference
lying in that according to Roemer the values are not necessarily
determined by a *linear* program. In Roemer's models the exploitation rate is well defined for each production process and

a more general version of the Fundamental Marxian Theorem can be proven within them. Nonetheless, the problem of translating values to prices —the so-called Transformation Problem— becomes an unsolvable problem and the Law of Value cannot be proven, which leads Roemer to conclude that Marx's theory of value is not a theory of commodity exchange. In Roemer's models, as well in the one built by Morishima on Von Neumann's, the role of the exploitation of workers, as a condition for the growth or reproduction of the economy, is well clarified. The models of Roemer are fairly general but they are still based upon the idealizing assumption that labor is homogeneous. We shall see in subsequent chapters how can we get rid of this assumption as well.

4.3.3.3 THE DIALECTICAL METHOD IN AXIOMATIC SYSTEMS

Once we have had the opportunity to observe the dialectical method operating in the sphere of axiomatic systems through some examples, it will be profitable to discuss in general its peculiar way of functioning in these cases, trying at the same time to eliminate some confusions in the terms in which it is often described, i.e. terms like 'rising to the concrete', 'idealization', 'concretization', 'general', 'particular' and other related terms. Throughout this work of elucidation I will try to reach a unitary and articulated vision of the dialectical method as applied to axiomatic theories.

We saw that according to Nowak the method that proceeds from the abstract to the concrete is none other than what he calls the "method of idealization", and that he characterizes this method as the one "which is able to reveal inner connections". Nevertheless, if by 'revealing inner connections' is understood the discovery of regularities capable of being expressed as scientific laws, then a distinction must be made between the method of passage to the concrete and that of idealization. Restricting ourselves to the domain of scientific theories that can be formulated by the definition of a set-theoretic predicate, it seems to me

that the primary sense of the term 'rising to the concrete' must
be identified with the procedure that consists in the construction of determined models of given theories in order to represent determined real concrete situations. By 'determined model'
I mean a model whose relations posses certain fixed values and
whose constants, in particular, have been assigned certain values,
i.e. I mean *one* model strictly speaking and not a *class* of models.
This term is exemplified here with model $\mathfrak{P}$ presented above,
since all its parameters are perfectly determined; but, if instead
of that, we had parameters like r_0, v_0 and g undetermined, then
we would rather have a *family* of models and not just one. When
I say that the model represents determined real concrete situations, what I mean is that his axioms are true of such situations
and so the values of the parameters predicted by the model have
a pragmatically acceptable degree of approximation with respect
to the corresponding available empirical data about those real
situations. Notice that, in this sense, "concrete" models are models that have instances in the real world.

Understanding primarily in this way the term 'rising to the
concrete', it is possible to observe other processes analogous to
the one described by this term. One of them is the construction
of a singular model which does not represent any real situation.
This method is useful, since it fulfills the purpose of building easily computable counterexamples (when they exist) to particular
theses that someone wants to disprove within the class of models
to which the mentioned model belongs. Yet the resulting models are not "concrete" but rather idealized, and so the process
is not exactly one of "rising to the concrete", but rather one of
deisolation and articulation of separated notions into a totality-
of-thought. Another, epistemologically more important process
consists in the construction of a class of models of the theory, by
means of a general description of the models in the same class,
that leaves at least some of the parameters of the class unspeci-
fied. The case can also obtain that no element of such class comes
to represent real situations, or that it represents them only in
a very rough approximated way. The procedure of idealization

—according to the present view— consists of the production of a class of models none of which constitutes a pragmatically acceptable approximation to the class of real situations being theorized about, precisely because some of the axioms defining such class are false in such situations. Hence, we can see that against what Nowak claims the method of idealization does not coincide with the one that moves from the abstract to the concrete, although idealization is indeed a case of deisolation and articulation of notions.

There is no doubt that at least in some cases the creation of a scientific theory begins with the production of idealized structures which become afterward a particular class of models of the theory, once the theory grows and its assumptions are made explicit. What Nowak calls 'concretization' is precisely the procedure which, taking as a point of departure a family of idealized structures, relaxes the assumptions defining the same structures by way of getting rid of false assumptions, thus obtaining a family of structures in which the fundamental law(s) which characterizes the original structure still hold. What Nowak calls 'concretization' is, therefore, a generalization process as well as a construction process of more realistic structures. It is a generalization process because the original idealized structures become a particular case of a wider class of structures, and it is also a construction process of more realistic structures because the elimination of the false assumptions defining the original structures gives rise to models in which every fundamental previous law still holds but which no longer depend on those false assumptions. The construction of these more realistic structures is, of course, a case of passage from the abstract to the concrete in the sense already explained, since it is not just a case of model construction out of abstract determinations, but one in which there is a tendency toward building a model with instances. Hence, it is seen that at least in the case of axiomatically defined models the dialectical method consists in a constructivistic procedure of deisolation and articulation of concepts, moved by a drive toward more concrete (i.e. instantiated) structures.

Thus, contrary to what Mäki (1991) suggests, it is not the same to concretize than to deisolate determinations. Also Hegel and Marx speak sometimes as if this were so. In the case of Hegel perhaps deisolating coincides with concretization, given the kind of concepts Hegel is using in the logic, where he is dealing with categories that have instantiations. In this sense, it would be more accurate to say that such concepts are concrete through and through. But if in the construction of a model one starts with idealized notions, the resulting model itself will never be concrete, inheriting the idealized character of the notions with which it was built, no matter how tightly articulated it may be. The prototype built by Marx, even though it is a unity of diverse determinations, still is too idealized to have instances. Hence, more than a mere articulation of isolated concepts into a totality is required in order to concretize. What is required is that the concepts have instances in the real world. An economic discourse built up exclusively with such kind of concepts I shall call "concrete"; an economic discourse that contains idealized concepts (like 'convex technology', 'continuous preferences', and so on) I will call 'idealized', and the economy it describes a *Meinongian* economy, because it is a nonexistent object. As a matter of fact, Marx's prototype is a Meinongian economy which fails to deisolate properly the concept of live labor, failing to connect that concept with the other concepts of the theory, his prototype being in fact the most idealized mathematical structure that can be built in economic theory. On the other hand, his fear to connect such concept with the determinations of capital is rather strange, and seems to have arisen from a misunderstanding. Having live labor in a given economic theory as labor-power, deeply connected with the determinations of capital, does not commit anybody to believe that human beings are *nothing but* a moment of capital. To say this is still to be playing with the idea of an Idea that determines itself. Dialectic in the form I have tried to explain only finds *conceptual* connections and only sublates *concepts*, not people. In the next chapter I will try to overcome the "contradictoriness" of Marx's concept of live labor through a new concept of abstract labor.

Chapter 5

ABSTRACT LABOR

As we saw in chapter 2, all well known mathematical formulations of MTV[1] are based upon Marx's assumption that labor in a capitalist economy is homogeneous, and so that the value of the goods it produces can be determined in terms of its mean or minimal temporal duration. In chapter 1 we saw Marx pretending that therefore value is an object that can be considered in the pure sphere of production, "independently of its form of manifestation" in the sphere of exchange. My point of departure in this chapter is the fact that Marx's methodological decision to sever value from exchange is in fact a dialectical mistake, that in effect value cannot be defined independently of the market concept. The argument to show this is simple and conclusive: the attempt to define value in that way inevitably leads to the transformation problem, a problem that according to Theorem 7 of chapter 1 is unsolvable unless we adopt the arbitrary assumption that all firms have the same value-composition of capital.

Marx himself provides in C, and perhaps more clearly in *A Contribution to a Critique of Political Economy*, elements to overcome the contradictoriness of his value concept. Recall that in C Marx almost said that the market is the social process, "that goes behind the back of the producers" that in effect reduces heterogeneous labors to a common unit. Even though in the *Contribution to a Critique of Political Economy* Marx seems to have already maintained his final view that value is determined in the sphere of production, from a certain point up in that work he begins to talk as if he were attributing a role to the market in that process. He says in part 1, chapter 1 of that work, for instance, the following:

> As exchange-values of different magnitudes they [i.e. commodities of different types] represent larger or smaller portions, larger or smaller amounts of simple, homogeneous, abstract general labour, which is the substance of exchange value.[2]

This passage sounds much the same as those passages in C where Marx defended the market-independent view of value. Nevertheless, as A Contribution progresses Marx begins to insist that in a commodity economy

> the labor of different persons is equated and treated as universal labour only by bringing one use-value into relation with another one in the guise of exchange-value.[3]

Or that in the exchange process

> universal social labor is [...] not a ready-made prerequisite but an emerging result.[4]

Naturally, putting together the first quotation with these last two a perplexity cannot but arise:

> [...] a new difficulty arises: on the one hand, commodities must enter the exchange process as materialised universal labour-time, on the other hand, the labour-time of individuals becomes materialised universal labour-time only as the result of the exchange process.[5]

It is pretty clear me that the contents of this last quotation is the manifestation of an ambiguity which is present both in Capital and in A Contribution. In both works Marx starts by describing the "substance" of value as homogeneous labor, which in addition he seems to identify with abstract labor and simple or unskilled labor. The impression that one gets reading the first sections of such books is that the "substance" of value can be defined almost in purely technological terms that can be applied to economies which are not necessarily capitalist, since Marx seems to characterize this "substance" —as we saw— as "an expenditure of human labour-power, in the physiological sense". Despite Marx's allegations that

> it is only the expression of equivalence between different sorts of commodities which brings to view the specific character of value-creating labour, by actually reducing the different kinds of labour embedded in the different kinds of commodity to their common quality of being human labour in general,[6]

one always gets the distinctive feeling that the so-called "form of value" is just a sociological appendage which is entirely irrelevant for the quantitative determination of value in terms of time socially necessary for the production of a commodity. This feeling is reinforced by Marx when he writes, in the same section, the following:

> Whether the coat is expressed as the equivalent and the linen as the relative value, or, inversely, the linen is expressed as equivalent and the coat as relative value, the magnitude of the coat's value is determined, as ever, by the labour-time necessary for its production, independently of its value-form.

The fact that Marx is so ambiguous, or even inconsistent, in dealing with the concept of abstract labor, does not mean that his theory cannot be reconstructed as based on the concept of abstract labor understood in a different way, in a way that takes the market process into account. The first economist in attempting this task was Isaak Illich Rubin in his *Essays on Marx's Theory of Value*, published for the first time in the Soviet Union in the mid twenties.[7] Rubin distinguishes three concepts of labor in Marx's theory, namely, (1) *physiologically equal* labor, (2) *socially equalized* labor, and (3) *abstract* or *abstract-universal* labor, i.e. "socially equalized labor in the specific form which it acquires in a commodity economy".[8] Physiologically equal or homogeneous labor is labor insofar as it is just an average productive expenditure of human energy in any form. Socially equalized labor is labor equalized not just as homogeneous labor, but rather as compared by some social factor or process; for instance, Rubin points out that in a socialist commune the different types of labor can be compared by a specific organ of the commune for the purpose of accounting and distribution of labor.[9] Labor compared in this way is *not* abstract labor. Abstract labor is labor compared through the market in a capitalist economy, where the firms are formally independent producers which are interconnected by the market (and not, say, by a central organ in charge of assigning production quotas to each firm). The category of abstract labor is historical and relative to the capitalist mode of production. It expresses the fact that in a commodity economy the different types of labor are compared and reduced to a common "substance" precisely in the process of commodity exchange, where the comparison of the different products induces a comparison of the corresponding labors that produced them. This is the form in which the social comparison of the different types of labor takes place in a capitalist economy.

The question that arises now is whether MTV, reconstructed on the basis of the concept of abstract labor as it has just been characterized, can be formulated in precise mathematical terms. The first effort in this direction was done by Professor Ulrich Krause at the end of the seventies in one book and two papers.[10] Roughly speaking, Krause was able to show that in economic systems that have certain special properties (that we shall see in the next section), the exchange of commodities induces a reduction of the different types of labor to a common measure, a reduction that can be represented by means of a linear functional defined on the set of all labor expenditures, understood as a subset of a linear space. Representing the labor-power applied in the production processes by means of n-dimensional vectors, as indicated in chapter 1, §3, the idea is to show that the exchange relations among the corresponding products do in fact induce an ordering among such vectors that can be represented by a linear functional. Assuming that certain special conditions are satisfied by the technology, this is precisely what Krause did, opening in this way the door to a mathematical treatment of the market-dependent view of value.

5.1 KRAUSE'S TREATMENT OF ABSTRACT LABOR

His most general treatment of abstract labor Krause gives in his 1980 paper "Abstract Labour in General Joint Systems". In this paper Krause introduces an economy that he labels a 'general joint system'. A *general joint system* is a triple of nonnegative matrices $\langle \mathbf{A}, \mathbf{B}, \mathbf{L} \rangle$ such that $\mathbf{A}$ is the $m \times l$ matrix of joint material inputs, $\mathbf{B}$ is the $m \times l$ matrix of joint material outputs, and $\mathbf{L}$ is the $n \times l$ matrix of joint labor inputs. The matrix of joint net outputs is defined as the difference $\mathbf{C} = \mathbf{B} - \mathbf{A}$ and a price system is a positive m vector $\mathbf{p} = [p_1 \cdots p_m]$. In the notation of chapter 1, $\mathbf{A}$ is a matrix of the form

$$\mathbf{A} = \begin{bmatrix} \underline{x}_{11} & \cdots & \underline{x}_{1l} \\ \vdots & & \vdots \\ \underline{x}_{m1} & \cdots & \underline{x}_{ml} \end{bmatrix}.$$

The columns of this matrix can be seen as the transposes of input vectors of processes $\tilde{\mathbf{x}}_1, \tilde{\mathbf{x}}_2, \ldots, \tilde{\mathbf{x}}_l$, which are all the processes in the economy. $\mathbf{B}$ is just the matrix whose columns are the transposes of the output vectors of these same processes:

$$\mathbf{B} = \begin{bmatrix} \bar{x}_{11} & \cdots & \bar{x}_{1l} \\ \vdots & & \vdots \\ \bar{x}_{m1} & \cdots & \bar{x}_{ml} \end{bmatrix}.$$

An *activation* or *state* of the joint system $\langle \mathbf{A}, \mathbf{B}, \mathbf{L} \rangle$ is an $l \times 1$ vector $\mathbf{s} \geqq \mathbf{0}$. This vector may be thought as indicating a given intensity of production of the economy; for instance, if $\mathbf{s}$ is the vector $[1 \cdots 1]$, the product

$$\mathbf{Bs} = \begin{bmatrix} \bar{x}_{11} + \cdots + \bar{x}_{1l} \\ \vdots \\ \bar{x}_{m1} + \cdots + \bar{x}_{ml} \end{bmatrix}$$

indicates the entire amounts of goods produced in the economy, their production requiring the consumption of inputs

$$\mathbf{Bs} = \begin{bmatrix} \underline{x}_{11} + \cdots + \underline{x}_{1l} \\ \vdots \\ \underline{x}_{m1} + \cdots + \underline{x}_{ml} \end{bmatrix}$$

and labor inputs

$$\mathbf{Ls} = \begin{bmatrix} x_{11} + \cdots + x_{1l} \\ \vdots \\ x_{n1} + \cdots + x_{nl} \end{bmatrix}.$$

At this intensity, the economy is producing the following vector of net outputs:

$$\mathbf{Cs} = \begin{bmatrix} \hat{x}_{11} + \cdots + \hat{x}_{1l} \\ \vdots \\ \hat{x}_{m1} + \cdots + \hat{x}_{ml} \end{bmatrix}.$$

Hence, the set of all possible net outputs is just the cone $\mathcal{C} = \{\mathbf{Cs} \mid \mathbf{s} \geq \mathbf{0}\}$, whereas the set of all possible expenditures of labor is the cone $\mathcal{L} = \{\mathbf{Ls} \mid \mathbf{s} \geq \mathbf{0}\}$.

According to Krause, "abstract labor means labor homogenized via the market by the exchange of products of labor". This homogenization of labor can be elucidated observing that *any* price system $\mathbf{p} \in \mathbb{R}^m_+$ induces a binary relation $\succsim$ over the cone $\mathcal{L}$, as follows. For any $\mathbf{x}, \mathbf{y} \in \mathcal{L}$, define $\succsim$ by the condition:

$$\mathbf{x} \succsim \mathbf{y} \quad \text{iff} \quad \mathbf{p}\hat{\mathbf{x}} \geq \mathbf{p}\hat{\mathbf{y}}.$$

Now, even though a relation like this is defined for every price system, not all price systems define the "right" relation between labor vectors. For it would seem that if the productive process $\tilde{\mathbf{x}}$ spends more or the same amount of labor-power than process $\tilde{\mathbf{y}}$ (i.e. $\mathbf{x} \geq \mathbf{y}$), then the price of the net output of $\tilde{\mathbf{x}}$ should be no less than the price of the net output of $\tilde{\mathbf{y}}$ ($\mathbf{p}\hat{\mathbf{x}} \geq \mathbf{p}\hat{\mathbf{y}}$). A price system $\mathbf{p}$ with this property is called 'admissible' by Krause, but I will call it 'valid' (*gültig*) from now on.

While exchange, which is the homogenization of products through money, can be described by prices collected in a price system $\mathbf{p}$, the homogenization of labor can be described by reduction coefficients, collected in a reduction $\mathbf{r}$, which is a positive $n \times 1$ vector. Krause defines an 'admissible' reduction as a reduction $\mathbf{r}$ that satisfies the condition:

$$\text{If } \hat{\mathbf{x}} \geq \hat{\mathbf{y}} \text{ then } \mathbf{rx} \geq \mathbf{ry}.$$

Krause then proves that any binary relation $\succsim$ induced by an admissible (that is, valid) price system is represented by an admissible reduction and that to any admissible reduction of labor there corresponds a valid price system. Hence, the price systems that define the "right" relation among vectors of labor inputs are precisely those that are valid. Abstract labor is a structure $\langle \mathcal{L}, \succsim \rangle$, where the relation $\succsim$ is induced by a valid price. Hence, in terms of the representation theory discussed in chapter 3, if $\succsim$ is induced by a valid price $\mathbf{p}$, an (admissible) reduction $\mathbf{r}$ corresponding to price $\mathbf{p}$ is just a representation of $\succsim$; i.e., for every $\mathbf{x}, \mathbf{y} \in \mathcal{L}$:

$$\mathbf{x} \succsim \mathbf{y} \quad \text{iff} \quad \mathbf{rx} \succsim \mathbf{ry}.$$

The labor-value of net output $\hat{\mathbf{x}} \in \mathcal{C}$, $\lambda(\hat{\mathbf{x}})$, is defined as the number $\mathbf{rx}$, and so labor-value is nothing but a representation of abstract labor in the sense of chapter 3 (notice, however, that it is not a *fundamental* measurement, since the structure $\langle \mathcal{L}, \succsim \rangle$ is not ontological).

Regarding the *existence* of valid prices (or admissible reductions), Krause proves that for given matrices $\mathbf{A}$ and $\mathbf{B}$ of inputs and ouputs there exists a semipositive price system (and a corresponding semipositive admissible reduction) for any $\mathbf{L}$ iff all commodities are separately producible, i.e. if for each bundle of commodities $\mathbf{b}$ there is an activation $\mathbf{s}$ of the joint system such that $\mathbf{b} = \mathbf{Cs}$. Krause also proves that there is a positive price (and a positive admissible reduction) for every $\mathbf{L}$ iff the previous condition is satisfied and in addition all processes are indispensable, i.e. if whenever the net output $\mathbf{Cs}$ is semipositive, the activation $\mathbf{s}$

is positive. Unfortunately, the conditions of separate producibility and indispensability are unduly restrictive.

The supposition that any good be separately producible is restrictive because it requires the existence of a labor process in $P = \{[\mathbf{Ls}, \mathbf{As}, \mathbf{Bs}] \mid \mathbf{s} \geq \mathbf{0}\}$ whose net output is a vector that has zeros everywhere except at a specified place. Since P is a cone, the condition can be formulated as follows:

(S9) $\forall i = 1, \ldots, m$: *there is a* $\tilde{\mathbf{x}} \in P$ *such that the ith entry of* $\hat{\mathbf{x}}$ *is one, and all the other entries are zero.*

(S9) holds in Leontief technologies but fails in some joint systems. For instance, it fails to hold in systems where two wage or luxury goods are jointly produced. For let the wage or luxury goods i and j $(k + 1 \leq i, j \leq m)$ be always jointly produced. Then, no matter in which process $\tilde{\mathbf{x}}$ they appear as outputs $(\overline{\mathbf{x}})$, since the entries i and j are always zero in $\underline{\mathbf{x}}$, the same places in the net output vector $\hat{\mathbf{x}} = \overline{\mathbf{x}} - \underline{\mathbf{x}}$ are also nonzero, and so none of the goods i and j can be separately produced. Hence, (S9) implies some form of no joint production.

The condition that all processes be indispensable is also restrictive because it requires that in order to produce any semipositive net output all processes of the economy be activated. But there is no reason why this should be so in general. Consider again the Leontief economy and let $\mathbf{e}_i$ be the m column vector that has zeros everywhere except at the ith place $(k + 1 \leq i \leq m)$, where it has a one, so that $\mathbf{e}_i$ represents one unit of a wage or luxury good. In order to produce precisely the bundle $\mathbf{e}_i$, the vector $\mathbf{s}_i$ of capital goods must be produced in such a way as to satisfy the equation[11]

$$\mathbf{As}_i + \underline{\mathbf{x}}_i^{\mathrm{T}} = \mathbf{s}_i.$$

Let $\overline{\mathbf{x}}^{\mathrm{T}} = \mathbf{s}_i + \mathbf{e}_i$ and think of the vector $\overline{\mathbf{x}}^{\mathrm{T}}$ as an activation of the joint system. Notice that this activation has some zero entries, in the places corresponding to the wage or luxury goods other than

i, because s_i has zeros at all places corresponding to non-capital goods. Then we have

$$\underline{\mathbf{x}}^T = \mathbf{A}\overline{\mathbf{x}}^T = \mathbf{A}s_i + \mathbf{A}e_i = [s_i - \underline{\mathbf{x}}_i^T] + \underline{\mathbf{x}}_i^T = s_i.$$

Hence,

$$\hat{\mathbf{x}}^T = \overline{\mathbf{x}}^T - \underline{\mathbf{x}}^T = [s_i + e_i] - s_i = e_i.$$

This shows that there is an activation that has some zeros and yet such that the net output is semipositive. Thus, the condition of indispensability, which can be formulated as

(S10) $\forall \mathbf{x} \in P$: *if* $\hat{\mathbf{x}} \geq 0$ *then* $\overline{\mathbf{x}} > 0$

is violated for instance in a Leontief technology.

　　Using (S9), (S10) and the Farkas-Minkowski Lemma, Krause proved the existence of a reduction, i.e. a positive vector $\mathbf{r}$ such that

$$\mathbf{rx} \geq \mathbf{ry} \qquad \text{iff} \qquad \mathbf{x} \succsim \mathbf{y},$$

for any labor input vectors $\mathbf{x}, \mathbf{y} \in \mathbf{L}$, for given matrices $\mathbf{A}$ and $\mathbf{B}$ and for *every* $\mathbf{L}$. This result is in itself very important but unduly strong. What we require is not a proof of the existence of valid prices and admissible reductions for given matrices $\mathbf{A}$ and $\mathbf{B}$ and *every* $\mathbf{L}$, but rather for every joint system that satisfies certain reasonable conditions. It is clear that the existence of a measurement of abstract labor cannot depend on assumptions as arbitrary as (S9) and (S10). In other words, *the basic problem in the foundations of* MTV *is to prove the existence of a cardinal measurement of abstract labor, without assuming conditions (S9) and (S10).* I shall tackle this problem in the next section.

5.2 THE CONCEPT OF ABSTRACT LABOR

Imagine a capitalist market economy in which there are l independent producers (capitalist firms). At the beginning of an economic cycle, each of these producers choses the production plan that he believes will yield the maximum profit for him. A production plan is a production process that has certain properties that will be formulated when the concept of a productive structure is introduced below. Recall that if m goods are being produced in the economy, and there are n different types of concrete labors, then we can represent a production process by means of a vector of the form $\tilde{\mathbf{x}} = [\mathbf{x}, \underline{\mathbf{x}}, \overline{\mathbf{x}}]$, where $\mathbf{x}$ is a nonnegative n vector whose ith component represents the amount of concrete labor of type i ($1 \leq i \leq n$) expended in the process $\tilde{\mathbf{x}}$, $\underline{\mathbf{x}}$ is a nonnegative m vector whose ith ($1 \leq i \leq m$) component represents the amount of goods of type i employed as means of production in $\tilde{\mathbf{x}}$, and $\overline{\mathbf{x}}$ is also a nonnegative m vector whose ith component represents the amount of goods of type i ($1 \leq i \leq m$) produced in $\tilde{\mathbf{x}}$. The vector $\mathbf{x}$ is called the vector of labor inputs, $\underline{\mathbf{x}}$ is the input vector, and $\overline{\mathbf{x}}$ is called the output vector. Notice that by definition the null vector $\tilde{\mathbf{0}}$ is a production process, but at any rate it follows that any production process $\tilde{\mathbf{x}}$ is nonnegative and has $2m + n$ components. It will be convenient also to have a separate notation for the net output of process $\tilde{\mathbf{x}}$, i.e. for the difference $\overline{\mathbf{x}} - \underline{\mathbf{x}}$; this shall be denoted (as before) as $\hat{\mathbf{x}}$. In order to represent the production plans of the firms, a set of production processes must possess the following additional property, namely, that each type of concrete labor be utilized at least in one of these plans and also that each type of good be either produced or used as a means of production in some production process; this property will be called 'nontriviality'. The former concepts are formally introduced in the following definition.

DEFINITION 1: A *production process* is a vector $[\mathbf{x}, \underline{\mathbf{x}}, \overline{\mathbf{x}}]$ in the linear space $\mathbb{R}^{2m+n}$, where $m, n \geq 1$, such that the n vector $\mathbf{x}$, as well as the m vectors $\underline{\mathbf{x}}$ and $\overline{\mathbf{x}}$, are all nonnegative. The vector $\mathbf{x}$ is called the *vector of labor inputs*, $\underline{\mathbf{x}}$ is called the *vector of means of production* or *input vector*, and $\overline{\mathbf{x}}$ is the *product vector* or *output vector*. The difference $\hat{\mathbf{x}} = \overline{\mathbf{x}} - \underline{\mathbf{x}}$ is called the *net output* of $\tilde{\mathbf{x}}$. A set Q of production processes is called *nontrivial* if for each i ($i = 1, ..., n$) there is a vector $\tilde{\mathbf{x}} \in Q$ such that the ith component x_i of the vector of labor inputs $\mathbf{x}$ of $\tilde{\mathbf{x}}$ is positive, and for each j ($j = 1, ..., m$) there is a vector $\tilde{\mathbf{x}} \in Q$ such that either the jth component $\underline{x}_j$ of the input vector of $\tilde{\mathbf{x}}$ is positive, or the jth component $\overline{x}_j$ of the output vector of $\tilde{\mathbf{x}}$ is positive.

Assuming that there is one price for each commodity, the price system is a nonnegative m vector $\mathbf{p} = [p_1 \cdots p_m]$. According to general equilibrium theory, to each producer h ($h = 1, ..., l$) there corresponds a production set Y^h of possible production processes representing his limited technological knowledge, and his behavior consists of choosing a point $\tilde{\mathbf{x}}_h$ in Y^h that maximizes his profit given the price system $\mathbf{p}$. I am not so much concerned here with the accuracy of this description (none of the results established in this chapter depends logically on it) as I am with the fact that as a result of this, $l > 0$ (possibly equal) production plans are implemented in the economy and, in fact, the different types of labor are indirectly compared among themselves through the comparisons between net outputs effected by the market forces by means of the price system. What this means is that there is a relation $\succsim$ among the vectors of labor inputs $\mathbf{x}_1, ..., \mathbf{x}_l$ of the production plans $\tilde{\mathbf{x}}_1, ..., \tilde{\mathbf{x}}_l$ chosen by the firms, such that

$$\mathbf{x}_h \succsim \mathbf{x}_i \quad \text{iff} \quad \mathbf{p}\hat{\mathbf{x}}_h \geq \mathbf{p}\hat{\mathbf{x}}_i \qquad (h, i = 1, ..., l)$$

i.e. such that $\mathbf{x}_h \succsim \mathbf{x}_i$ (to be read as: "$\mathbf{x}_h$ represents at least as much social labor as $\mathbf{x}_i$") whenever the price of the net output produced by the labor expenditures represented by $\mathbf{x}_h$ is greater than the price of the net output produced by the labor expenditures represented by $\mathbf{x}_i$. From a logical point of view, $\succsim$ is a binary relation which is connected, reflexive and transitive on the set of labor input vectors. Moreover, defining $\sim$ as usual, namely, by the condition

$$\mathbf{x}_h \sim \mathbf{x}_i \quad \text{iff} \quad \mathbf{x}_h \succsim \mathbf{x}_i \quad \text{and} \quad \mathbf{x}_i \succsim \mathbf{x}_h,$$

we could also prove that $\sim$ is an equivalence relation over the same set. What this means is that *any* system of market prices does in fact induce a comparison of the different types of labor operating in the economy. The strict dominance relation $\succ$ is defined, as usual, by means of the condition

$$\mathbf{x}_h \succ \mathbf{x}_i \quad \text{iff} \quad \mathbf{x}_h \succsim \mathbf{x}_i \quad \text{and not} \quad \mathbf{x}_i \succsim \mathbf{x}_h.$$

Not every relation induced by a price over the labor inputs, however, counts as abstract labor. It seems intuitively clear that in order for a relation over a set of labor inputs to be thought as abstract labor, the exchange relation among their corresponding net outputs must be valid (*gültig*). The very minimum condition for validity is that exchanges be ruled out, in which net products that "contain" more concrete labor are traded for net products that "contain" less. This can be expressed symbolically by the two following conditions:

(C1) If the vector of labor inputs $\mathbf{x}_h$ of process $\tilde{\mathbf{x}}_h$ is equal to the labor vector $\mathbf{x}_i$ of process $\tilde{\mathbf{x}}_i$ (i.e. $\mathbf{x}_h = \mathbf{x}_i$), then $\mathbf{x}_h \sim \mathbf{x}_i$.

(C2) If some component of the vector of labor inputs $\mathbf{x}_h$ of process $\tilde{\mathbf{x}}_h$ is strictly greater than the corresponding component of the labor vector $\mathbf{x}_i$ of process $\tilde{\mathbf{x}}_i$, and no component of $\mathbf{x}_i$ is strictly greater than the corresponding component of $\mathbf{x}_h$ (i.e. $\mathbf{x}_h \geq \mathbf{x}_i$), then $\mathbf{x}_h \succ \mathbf{x}_i$.

What (C1) says is that each vector of labor expenditures represents a fixed quantity of social labor or, what is the same, that no labor vector represents more social labor than itself. (C2) asserts that if vector of labor inputs $\mathbf{x}_h$ contains at least as much labor of each type as labor vector $\mathbf{x}_i$, and in fact more concrete labor of at least one type than labor vector $\mathbf{x}_i$, then $\mathbf{x}_h$ represents more social labor than $\mathbf{x}_i$. It is easy to see that a necessary and sufficient condition for conditions (C1) and (C2) to hold is that the abstract labor relation be induced by a price which is admissible in Krause's sense. For reasons that were apparent in chapter 1, I prefer the term 'valid', a term which is precisely defined now.

DEFINITION 2: Let P be any set of production processes. A *price system* (or simply a *price*) for P is a positive m vector $\mathbf{p} = [p_1 \cdots p_n]$. If, in addition, $\mathbf{p}$ satisfies

$$\text{If} \quad \mathbf{x} \geqq \mathbf{y} \quad (\text{resp. } \geq), \quad \text{then} \quad \mathbf{p}\hat{\mathbf{x}} \geq \mathbf{p}\hat{\mathbf{y}} \quad (\text{resp. } >),$$

then the price $\mathbf{p}$ is called *valid* for P.

In terms of the concept of valid price system, the definition of the concept of abstract labor is straightforward and can be introduced at this point. According to Rubin (1972), the introduction of this concept is what distinguishes Marx's labor theory of value from that of Ricardo's. It will be obvious from the definition that abstract labor is a relation generated by a valid price on a system of production processes and that such relation actually represents an ordering of all the labor expenditures of the system.

DEFINITION 3: Let P be a set of production processes. The set $\mathcal{L} = \{\mathbf{x} : [\mathbf{x}, \underline{\mathbf{x}}, \overline{\mathbf{x}}]\}$ of all the vectors of labor inputs of processes in P is called the *set of labors* of P. If $\mathbf{p}$ is a valid price for P and $\succsim$ is the binary relation induced by $\mathbf{p}$ over $\mathcal{L}$, then $\succsim$ is called *abstract labor* and the structure $\mathfrak{L} = \langle \mathcal{L}, \succsim \rangle$ is called the *abstract labor structure corresponding to P and* $\mathbf{p}$.

This concept of abstract labor elucidates in precise terms Marx's suggestion that the reduction of the different types of concrete labors to a common measure is effected by a process that goes behind the backs of the producers. As I suggested in chapter 1, in a capitalist economy this process cannot be but the market. It is the market the process that reduces these labors to a common unit. Marx spoke of reducing all kinds of labor to simple unskilled labor, but there is no need to take this type of labor as the standard and in fact any other would do equally well. I shall discuss in the next section the conditions that guarantee the existence of valid price systems and therefore of abstract labor. My approach shall be different from that of Krause's. As we saw, he proved that a valid price system necessarily goes together with an admissible reduction; he then proceeded to show that for joint systems satisfying (S9) and (S10) both determinations exist. My strategy shall be instead to try to find conditions over production processes, wages and profit rates that guarantee the existence of the abstract labor relation. We know that to any valid price system there corresponds a nonnegative reduction, but we do not know whether this reduction is positive. In the final section of the present chapter I will show that every abstract labor relation can be represented or measured (in the sense of chapter 3) by a positive reduction.

5.3 THE EXISTENCE OF ABSTRACT LABOR

A question that naturally arises concerns the conditions under which a price system is valid in an economy, i.e. the conditions under which the price system actually induces the abstract labor relation. In order to discuss this problem, I will consider a joint system, i.e. the convex polyhedral cone of labor processes spanned by the production processes actually operated by the firms in the economy. I will introduce the concepts of wage system and profit rate. A wage system is an n vector $\mathbf{w}$ whose ith entry is a positive number which represents the hourly wage of workers of type i ($1 \leq i \leq n$). Thus, if $\mathbf{x}$ is the vector of labors of production process $\tilde{\mathbf{x}}$, then $\mathbf{wx}$ represents the monetary costs of labor power required to operate $\tilde{\mathbf{x}}$, i.e. what Marx called the "variable" capital (in terms of money) of production process $\tilde{\mathbf{x}}$. The profit rate of $\tilde{\mathbf{x}}$, on the other hand, is the number $\pi(\tilde{\mathbf{x}})$ that satisfies the following equation:

$$\mathbf{p\bar{x}} = [1 + \pi(\tilde{\mathbf{x}})](\mathbf{p\underline{x}} + \mathbf{wx}).$$

That is, the profit rate multiplied by the cost of production yields the net profit of process $\tilde{\mathbf{x}}$. More formally, these concepts can be introduced as follows.

DEFINITION 4: Let P be a set of production processes. A *wage system* for P is a positive n vector $\mathbf{w} = [w_1 \cdots w_n]$ of real numbers. If $\mathbf{w}$ is a wage system, the *profit rate* of a labor process $\tilde{\mathbf{x}}$ is the number $\pi(\tilde{\mathbf{x}})$ given by the equation

$$\pi(\tilde{\mathbf{x}}) = \frac{\mathbf{p\hat{x}} - \mathbf{wx}}{\mathbf{p\underline{x}} + \mathbf{wx}},$$

if $\mathbf{x}$ is nonnull, and zero otherwise.

Returning to the problem of the conditions under which a price system is valid in an economy, it can be shown that if the rate of profit is uniform in the economy, then the price $\mathbf{p}$ happens to be valid, provided that a certain normality assumption connecting expenditures of labor and means of production is granted.

Consider the cone P spanned by the actually operated processes in the economy, the elements of $P_0 = \{\tilde{x}_1, ..., \tilde{x}_l\}$. Our approach to the problem of the existence of valid prices for P and so for P_0, which is the set that mainly interests us, is closely linked to the existence of a uniform profit rate. We would like to prove that if in the economy determined by processes $\tilde{x}_1, ..., \tilde{x}_l$ labor is both productive and indispensable,[12] and the profit rate is uniform, then the price is valid. How can we do this?

It would seem that the clue lies in the concept of efficient process. A process $\tilde{x}$ is more efficient than a process $\tilde{y}$ iff $\tilde{x}$ produces more than $\tilde{y}$ using the same or less amounts of labor, or produces the same or more using less amounts of labor. More precisely, we shall follow the use of the term introduced by the following definition.

DEFINITION 5: Let $\tilde{x} = [x, \underline{x}, \overline{x}]$ and $\tilde{y} = [y, \underline{y}, \overline{y}]$ be two production processes. We say that $\tilde{x}$ is *more efficient than* $\tilde{y}$ iff the net output of $\tilde{x}$ is greater than the net output of $\tilde{y}$, even though $\tilde{x}$ does not expend more labor of any type than $\tilde{y}$, or the net output of $\tilde{x}$ is equal or greater than the net output of $\tilde{y}$ even though $\tilde{x}$ spends less labor of some type than $\tilde{y}$. In symbols:

$$\tilde{x}E\tilde{y} \quad \text{iff} \quad \text{either} \quad x \leq y \quad \text{and} \quad \hat{x} \geq \hat{y},$$
$$\text{or} \quad x < y \quad \text{and} \quad \hat{x} \geq \hat{y}$$

where '$\tilde{x}E\tilde{y}$' stands for '$\tilde{x}$ is more efficient than $\tilde{y}$'.

Indeed, if no process is more efficient than any other, then it follows that $x \leq y$ implies $\hat{x} \leq \hat{y}$ for arbitrary processes $\tilde{x}$ and $\tilde{y}$. Hence, in such a case the inequality $x \leq y$ implies $px \leq py$; i.e. it follows that p is valid. Thus, if we could show that processes with different degrees of efficiency cannot have the same rate of profit, that would be enough to establish what we want, namely, that prices that equalize profit rates are valid.

Unfortunately, this proposition is false. Consider an economy with two processes $\tilde{x} = [(1, 1), (3/2, 1), (5, 3/2)]$ and $\tilde{y} =$

[(2, 2), (1/2, 1/4), (3, 1/2)], a price system $\mathbf{p} = [1, 2]$ and a wage system $\mathbf{w} = [1/4, 1/4]$. We have in this economy $\hat{\mathbf{x}} = [7/2, 1/2]$, $\hat{\mathbf{y}} = [5/2, 1/4]$, and so we see at once that $\tilde{\mathbf{x}}$ is more efficient than $\tilde{\mathbf{y}}$ in the sense of Definition 5. Nevertheless, quick computations using Definition 4 show that $\pi(\tilde{\mathbf{x}}) = 1 = \pi(\tilde{\mathbf{y}})$.

An examination of the counterexample just given reveals a curious situation: How is it possible that the workers in process $\tilde{\mathbf{y}}$, working twice as much as the workers in process $\tilde{\mathbf{x}}$, spend about 3 times less means of production than those in process $\tilde{\mathbf{x}}$? Clearly, this would be the case only if either the workers in process $\tilde{\mathbf{x}}$ were wasting means of production, or the workers in process $\tilde{\mathbf{y}}$ were working too slowly, but according to Marx this situation cannot obtain in the valorization process, since in such a process "the time spent in production counts only in so far as it is socially necessary for the production of a use value".[13] According to Marx, this has various consequences, one of them being that

> all wasteful consumption of raw material or instruments of labor is strictly forbidden, because what is wasted in this way represents a superfluous expenditure of quantities of objectified labour, labour that does not count in the product or enter into its value.[14]

Another consequence is that

> the labour-power itself must be of normal effectiveness. In the trade in which it is being employed, it must possess the average skill, dexterity and speed prevalent in that trade [...and...] it must be expended with the average amount of exertion and the usual degree of intensity.[15]

It seems intuitively clear that the two former consequences preclude situations like the one of the former counterexample. Indeed, consider two workers of the same trade fulfilling the just given conditions, say two painters. Since none of them works faster than the other, and none of them wastes paint and brushes unnecessarily, it is utterly unreasonable to expect that any of them will spend more means of production per unit time than the other. Let us call 'normal' the labor-power satisfying this couple of conditions. From a quantitative point of view, we say that the labor power employed in two process $\tilde{x}$ and $\tilde{y}$ is normal if whenever $\mathbf{x} \geq \mathbf{y}$ it follows that $\underline{\mathbf{x}} \geq \underline{\mathbf{y}}$. More precisely, I introduce

DEFINITION 6: Let Q be a set of production processes. We say that the labor employed in the processes in Q is *normal* iff none of them is null and, for every $\tilde{\mathbf{x}}, \tilde{\mathbf{y}} \in Q$, $\mathbf{x} \geq \mathbf{y}$ ($\geq$) implies that $\underline{\mathbf{x}} \geq \underline{\mathbf{y}}$ ($\geq$). We shall also call the set Q itself *normal* when the labor employed in it is normal.

The following theorem, which answers the question that we have been considering, is an important consequence of a set of production processes' being normal.

THEOREM 1: *Let Q be a normal set of production processes. If the rate of profit is the same for every process in Q, under the price $\mathbf{p}$ and the wage system $\mathbf{w}$, then $\mathbf{p}$ is valid.*

Proof: Let $\tilde{\mathbf{x}}, \tilde{\mathbf{y}}$ be any two elements of Q and assume that the profit rate is uniform for all the elements of Q, so that $\pi(\tilde{\mathbf{x}}) = \pi(\tilde{\mathbf{y}})$. If $\mathbf{x} \leq \mathbf{y}$ then we have two cases: either (i) $\mathbf{x} = \mathbf{y}$ or (ii) $\mathbf{x} \leq \mathbf{y}$. If (i) is the case, since $\mathbf{w}$ is positive and $\mathbf{p}$ semipositive, $\mathbf{wx} = \mathbf{wy}$ and $\mathbf{p}\underline{\mathbf{x}} = \mathbf{p}\underline{\mathbf{y}}$. Hence, $\mathbf{wx} + \mathbf{p}\underline{\mathbf{x}} = \mathbf{wy} + \mathbf{p}\underline{\mathbf{y}}$ and so

$$\mathbf{p}\hat{\mathbf{x}} - \mathbf{wx} = \pi(\tilde{\mathbf{x}}) \cdot (\mathbf{wx} + \mathbf{p}\underline{\mathbf{x}}) = \pi(\tilde{\mathbf{y}})(\mathbf{wy} + \mathbf{p}\underline{\mathbf{y}}) = \mathbf{p}\hat{\mathbf{y}} - \mathbf{wy}.$$

It follows that $\mathbf{p}\hat{\mathbf{x}} = \mathbf{p}\hat{\mathbf{y}}$.

If (ii) is the case, then $\mathbf{wx} < \mathbf{wy}$ and $\mathbf{p}\underline{\mathbf{x}} \leq \mathbf{p}\underline{\mathbf{y}}$. Hence, $\mathbf{wx} + \mathbf{p}\underline{\mathbf{x}} < \mathbf{wy} + \mathbf{p}\underline{\mathbf{y}}$ and so

$$\mathbf{p}\hat{\mathbf{x}} - \mathbf{wx} = \pi(\tilde{\mathbf{x}}) \cdot (\mathbf{wx} + \mathbf{p}\underline{\mathbf{x}}) < \pi(\tilde{\mathbf{y}})(\mathbf{wy} + \mathbf{p}\underline{\mathbf{y}}) = \mathbf{p}\hat{\mathbf{y}} - \mathbf{wy}.$$

Adding $\mathbf{wx}$ at both sides of the extreme expressions, we get $\mathbf{p}\hat{\mathbf{x}} < \mathbf{p}\hat{\mathbf{y}} + (\mathbf{wx} - \mathbf{wy}) < \mathbf{p}\hat{\mathbf{y}}$, because $\mathbf{wx} - \mathbf{wy}$ is negative. $\square$

Theorem 1 shall be used as a lever to build a couple of interesting models of the theory in subsequent chapters. We shall always suppose that the aggregate technology possibility sets are normal, thus implying that in particular the cone spanned by the set P_0 of processes actually chosen by the firms is also normal. This will follow from the fact that such cone is in fact a subset of the aggregate technology possibility set and that any subset of a normal set is also normal (Theorem 2). We shall also exploit theorems 3 and 4.

THEOREM 2: *If the set of production processes Y is normal, and X is any nonempty subset of Y, then X is normal.*

Proof: Assume that Y is normal and that $X \subseteq Y$. Let $\tilde{\mathbf{x}}, \tilde{\mathbf{y}} \in X$, and suppose that $\mathbf{x} \geq \mathbf{y}$. Since $\tilde{\mathbf{x}}, \tilde{\mathbf{y}} \in Y$ also, it is immediate that $\underline{\mathbf{x}} \geq \underline{\mathbf{y}}$ and so X is normal. $\square$

THEOREM 3: *If P^+ is the positive hull spanned by $P_0 = \{\tilde{\mathbf{x}}_1, ..., \tilde{\mathbf{x}}_l\}$, and the profit rate r is the same for all the elements of P_0 under the price system $\mathbf{p}$, then it is also the same for all the elements of P^+.*

Proof: We can assume, without any loss of generality, that P_0 is linearly independent (otherwise, just drop some of its elements). If $\tilde{\mathbf{x}} \in P^+$ then there exist unique nonnegative numbers $\alpha_1, ..., \alpha_l$ such that $\tilde{\mathbf{x}} = \alpha_1 \tilde{\mathbf{x}}_1 + \cdots \alpha_l \tilde{\mathbf{x}}_l$. Thus, as a matter of fact, $\mathbf{x} = \sum_{h=1}^{l} \alpha_h \mathbf{x}_h$, $\underline{\mathbf{x}} = \sum_{h=1}^{l} \alpha_h \underline{\mathbf{x}}_h$, $\bar{\mathbf{x}} = \sum_{h=1}^{l} \alpha_h \bar{\mathbf{x}}_h$ and so $\hat{\mathbf{x}} = \bar{\mathbf{x}} - \mathbf{x} = \sum_{h=1}^{l} \alpha_h (\bar{\mathbf{x}}_h - \underline{\mathbf{x}}_h) = \sum_{h=1}^{l} \alpha_h \hat{\mathbf{x}}_h$. Hence,

$$\pi(\tilde{\mathbf{x}}) = \frac{\mathbf{p}(\sum_{h=1}^{l} \alpha_h \hat{\mathbf{x}}_h) - \mathbf{w}(\sum_{h=1}^{l} \alpha_h \mathbf{x}_h)}{\mathbf{p}(\sum_{h=1}^{l} \alpha_h \underline{\mathbf{x}}_h) + \mathbf{w}(\sum_{h=1}^{l} \alpha_h \mathbf{x}_h)}$$

$$= \frac{\alpha_1(\mathbf{p}\hat{\mathbf{x}}_1 - \mathbf{w}\mathbf{x}_1) + \cdots + \alpha_l(\mathbf{p}\hat{\mathbf{x}}_l - \mathbf{w}\mathbf{x}_l)}{\alpha_1(\mathbf{p}\underline{\mathbf{x}}_1 + \mathbf{w}\mathbf{x}_1) + \cdots + \alpha_l(\mathbf{p}\underline{\mathbf{x}}_l + \mathbf{w}\mathbf{x}_l)}$$

Thus, setting

$$\beta_h = \frac{\alpha_h(\mathbf{p}\underline{\mathbf{x}}_h + \mathbf{w}\mathbf{x}_h)}{\sum_{h=1}^{l} \alpha_h(\mathbf{p}\underline{\mathbf{x}}_h + \mathbf{w}\mathbf{x}_h)}$$

we have $\sum_{h=1}^{l} \beta_h = 1$ and get

$$\pi(\tilde{\mathbf{x}}) = \frac{\alpha_1 \beta_1 (\mathbf{p}\hat{\mathbf{x}}_1 - \mathbf{w}\mathbf{x}_1)}{\alpha_1 (\mathbf{p}\underline{\mathbf{x}}_1 + \mathbf{w}\mathbf{x}_1)} + \cdots + \frac{\alpha_l \beta_l (\mathbf{p}\hat{\mathbf{x}}_l - \mathbf{w}\mathbf{x}_l)}{\alpha_l (\mathbf{p}\underline{\mathbf{x}}_l + \mathbf{w}\mathbf{x}_l)}$$

$$= \sum_{h=1}^{l} \beta_h r$$

$$= r. \quad \square$$

THEOREM 4: *If P^+ is the positive hull spanned by $P_0 = \{\tilde{\mathbf{x}}_1, ..., \tilde{\mathbf{x}}_l\}$, the profit rate r is the same for all the elements of P_0 under price system* **p**, *and P^+ is normal, then* **p** *is valid for $P = P^+ \cup \{\tilde{\mathbf{0}}\}$.*

Proof: This follows immediately from Theorems 1 and 3. $\square$

Roughly speaking, a productive structure is the convex cone generated by the production processes chosen by the capitalist firms. The only reason why this cone is introduced is to make computations feasible; the reader may interpret this cone as representing the technology actually chosen by the firms. Additional axioms defining the concept of productive structure are that the set of production processes chosen by the firms is nontrivial and that no input vector of these processes is null, i.e. that every production process requires some positive amounts of means of production; to this condition will be added the assumption that labor is both productive and indispensable and that the cone is normal. It is easy to see that the former conditions imply that in every production plan positive amounts of goods are obtained out of positive amounts of means of production and concrete labors. The concept of a productive structure is formally introduced by the following definition.

DEFINITION 7: The structure $\mathfrak{P} = \langle P_0, P \rangle$ is a *productive structure* iff it satisfies the following axioms:

(P1) P_0 is a nontrivial finite set of labor processes

(P2) P is the convex cone generated by P_0

(P3) For every $\langle \mathbf{x}, \underline{\mathbf{x}}, \overline{\mathbf{x}} \rangle \in P_0$, $\mathbf{x} \geq 0$ and labor is productive and indispensable, i.e. the following implications hold:

$$\underline{\mathbf{x}} \geq \underline{0} \Rightarrow \mathbf{x} \geq 0 \Leftrightarrow \overline{\mathbf{x}} \geq \overline{0}$$

(P4) P is normal.

The properties of the production processes in P_0 expressed by (P3) also hold for all processes in the cone P. This is established as

THEOREM 5: *For every* $\langle \mathbf{x}, \underline{\mathbf{x}}, \overline{\mathbf{x}} \rangle \in P$,

$$\underline{\mathbf{x}} \geq 0 \Rightarrow \mathbf{x} \geq 0 \Leftrightarrow \overline{\mathbf{x}} \geq 0.$$

Proof: Let $\tilde{\mathbf{x}} = [\mathbf{x}, \underline{\mathbf{x}}, \overline{\mathbf{x}}]$ be any production process in P. Then there exist nonnegative $\alpha_1, \ldots, \alpha_l$ such that

$$\tilde{\mathbf{x}} = \alpha_1 \tilde{\mathbf{x}}_1 + \cdots + \alpha_l \tilde{\mathbf{x}}_l.$$

If $\mathbf{x} \geq 0$, then some of the $\alpha_1, \ldots, \alpha_l$, say $\alpha_1, \ldots, \alpha_k$ are positive. In such a case, since (by (P3))

$$\underline{\mathbf{x}}_i \geq \underline{0} \Rightarrow \mathbf{x}_i \geq 0 \Leftrightarrow \overline{\mathbf{x}}_i \geq \overline{0}$$

It follows that $\mathbf{x}_i \geq 0 \Leftrightarrow \overline{\mathbf{x}}_i \geq \overline{0}$, which implies that

$$\mathbf{x} = \sum_i^k \alpha_i \mathbf{x}_i \geq 0 \Leftrightarrow \overline{\mathbf{x}} = \sum_i^k \alpha_i \overline{\mathbf{x}}_i \geq \overline{0}$$

I shall conclude the present section by formulating the following methodologically useful theorem, which follows almost immediately from theorem 4 and 5.

THEOREM 6: *Let* $\mathfrak{P} = \langle P_0, P \rangle$ *be a productive structure,* $\mathcal{L}$ *the set of labors of P and* $\succsim$ *the relation over $\mathcal{L}$ generated by price* $\mathbf{p}$. *If the profit rate of the production processes in P_0 is nonnegative and uniform at prices* $\mathbf{p}$, *then* $\langle \mathcal{L}, \succsim \rangle$ *is an abstract labor structure.*

Theorem 6 is methodologically important because it shows that in order to establish the existence of abstract labor in a productive structure it is sufficient to show the existence of a profit equalizing equilibrium. This solves the dialectical contradiction, pointed out in previous chapters, between the observed tendency toward a uniform profit rate in a market economy and labor-value. In fact, according to Theorem 6 this tendency leads straightforwardly to abstract labor, to reduce all labors to a common standard.

Roemer (1980, 1981) has established the existence of a profit equalizing equilibrium for several models of MTV. These models are built upon the assumption that labor is homogeneous and do not depend on (P4), but they can be modified by introducing heterogeneous labor in such a way that the existence of the required profit equalizing equilibria is also provable within them. I will provide the details of this in chapters 7 and 8.

5.4 THE REPRESENTATION OF ABSTRACT LABOR

The present section is devoted to the main aim of the chapter, which is to prove the existence of a mathematical representation of abstract labor. We saw on chapter 3 what a mathematical representation is and, in particular, we introduced there the concept of a fundamental measurement. The measurement of abstract labor as provided in the present chapter is indeed a representation in that sense, but not a *fundamental* measurement, since the structure to be represented —the abstract labor structure— is not ontological but is already a mediated representation of concrete labor; this structure is itself a mathematical representation of an aspect of the production processes taking place in a market economy.

A mathematical representation of abstract labor is a function $\varphi: \mathcal{L} \rightarrow \mathbf{R}$ such that $\varphi(\mathbf{x}) \geq \varphi(\mathbf{y})$ whenever $\mathbf{x} \succsim \mathbf{y}$, for every $\mathbf{x}, \mathbf{y} \in \mathcal{L}$. As a matter of fact, there are quite a few of such representations. Just consider the function φ that assigns to each $\mathbf{x} \in \mathcal{L}$ the number $\mathbf{p}\hat{\mathbf{x}}$ (which is unique according to Definition

2). If $\psi: \mathbf{R} \to \mathbf{R}$ is any strictly increasing function, then the composition $\psi \circ \varphi$ is another representation of abstract labor. Representations of this type are called in the literature "ordinal" representations. This is not the type of representation with which the present chapter is concerned. This chapter is concerned with a very specific type of what is called in the literature a "cardinal" representation. More precisely, the representation we are looking for is a very specific type of what is called in Krantz, *et al.* (1971) an additive conjoint measurement. An additive conjoint measurement of a binary relation R is a family of real-valued functions $\{\varphi_i\}_{i \in N}$, where $N = \{1, ..., n\}$, such that

$$\sum_{i \in N} \varphi_i(x_i) \geq \sum_{i \in N} \varphi_i(y_i) \quad \text{iff} \quad [x_1 \cdots x_n]R[y_1 \cdots y_n],$$

where $[x_1 \cdots x_n]$ and $[y_1 \cdots y_n]$ are any two vectors in the field of R. In the particular case of abstract labor it can be shown that functions φ_i can be found that act upon the vectors that take as arguments as the tensor "taking the inner product with a fixed positive n vector", i.e. if $[x_1 \cdots x_n]$ is a vector in $\mathcal{L}$, then there is a positive vector $\mathbf{r} = [r_1 \cdots r_n]$ such that $\varphi_i(x_i) = r_i x_i$, and so

$$\sum_{i \in N} \varphi_i(x_i) = \mathbf{r} \mathbf{x}.$$

If we call any vector $\mathbf{r}$ having the former characteristics a *reduction*, then the aim of the Representation Theorem can be described as that of establishing the existence of a reduction. The use of this term is justified because it effects a reduction of the different heterogeneous labors to a common measure (which by definition is none other than labor-value).

To show that there is a reduction for a particular abstract labor structure is tantamount to showing that the market actually assigns a specific weight to each and every type of labor, even though this assignment may not be unique. Perhaps a physical analogy will be useful here. Imagine a multiarm star-shaped balance whose arms are of equal length and each pair of which forms a constant angle, so that if the balance is hanged from the center of the star the arms are balanced. Supposing that there are l production processes in the economy, we compare each process with an arm of one such balance with l arms, assuming that each arm has n numbered marks where lead spheres can be hanged. The ith mark ($i = 1, ..., n$) in arm j ($j = 1, ..., l$) indicates the place where the ith sphere must be hanged, and the distance from this mark to the center of the star is the amount of concrete labor of type i expended in the jth process. The problem then is to find l equal sets of n spheres such that, hanging on each arm the ith sphere on the mark numbered i, the arms of the balance are again in equilibrium, i.e. all of them still are parallel to the floor. These spheres are the physical analogues of the weights that the market assigns to each type of concrete labor, and the fact that n spheres can always be found for equilibrium means that the value of each net output is decomposed as a contribution of *all* the concrete labors that gave rise to it. Notice that the problem of the representation of abstract labor can always be posed in these terms, because given the cone spanned by any finite set of production processes the assumption of constant returns to scale guarantees that multiples of the same can always be taken such that the price of the net output is the same for them all. Actually, this procedure will be followed in the proof of the Representation Theorem below.

Some reader might wonder whether the weights assigned by the market to the different heterogeneous labors are not just the hourly wages assigned to them. Actually, we can build an example where this is so. Consider a productive structure in which material inputs are a linear function of labor inputs. More precisely, for every type of good j ($j = 1, .., m$), there are nonnegative scalars $\mu_{1j}, ..., \mu_{nj}$ such that, for every $\tilde{\mathbf{x}}$ in P_0 with vector of labor inputs $\mathbf{x} = [x_1 \cdots x_n]$ and vector of inputs $\underline{\mathbf{x}} = [\underline{x}_1 \cdots \underline{x}_m]$,

$$\underline{x}_j = \sum_{i=1}^{n} \mu_{ij} x_i$$

where $\underline{x}_j$ is the jth component of $\underline{x}$ and x_i is the ith component of x. Let p be a price system and let M be the matrix

$$M = \begin{bmatrix} \mu_{11} & \mu_{12} & \cdots & \mu_{1n} \\ \mu_{21} & \mu_{22} & \cdots & \mu_{2n} \\ \vdots & \vdots & \ddots & \vdots \\ \mu_{m1} & \mu_{m2} & \cdots & \mu_{mn} \end{bmatrix}.$$

Then it is easy to see that the vector r given by

$$r = (1 + r)w + rpM$$

is a reduction. In fact, if $x \in A$,

$$\begin{aligned} rx &= (1 + r)wx + rpMx^T \\ &= (1 + r)wx + rp\underline{x} \\ &= (1 + r)(p\underline{x} + wx) - p\underline{x} \\ &= p\bar{x} - p\underline{x} \\ &= p\hat{x}. \end{aligned}$$

It is easy to see that when the rate of profit is zero, $r = w$, i.e. the wage vector is a reduction of labor.

Nevertheless, it can also be shown that the wage vector is not always a reduction. Consider the productive structure spanned by the set $P_0 = \{\tilde{x}, \tilde{y}\}$, where $\tilde{x} = [(1, 0), (2, 20), (1, 45)]$ and $\tilde{y} = [(0, 1), (2, 14), (1, 35)]$. Then

$$M = \begin{bmatrix} 2 & 20 \\ 2 & 14 \end{bmatrix}.$$

Set $w = [1, 2]$, $p = [1, 1]$. It follows that $\pi(\tilde{x}) = \pi(\tilde{y}) = 1$ and trivial computations show that in spite of the fact that by Theorem 4 p is admissible, and $p\hat{x} = 24 > 20 = p\hat{y}$, still $wx = 1 < 2 = wy$.

Very much in the style of chapter 3, I shall proceed now to provide a representation theorem for abstract labor structures.

I will prove that there is at least one positive reduction and that this reduction is unique up to affine transformations. The theorem does not depend logically upon Axiom (P4) but this does not matter, since the aim of that axiom is only to prove that the competition of capitals leads to abstract labor whenever the processes in the economy are normal. At any rate we have the following central result.

THEOREM 7: (Representation Theorem). *Let* $\langle \mathcal{L}, \succeq \rangle$ *be the abstract labor structure corresponding to the set of production processes P and the price system* **p**, *and suppose that P is a convex polyhedral cone generated by a set* P_0 *that satisfies (P1) and (P3). Then there exists a positive vector* **r** *such that, for every* $x, y \in \mathcal{L}$,

$$\mathbf{r}x \geq \mathbf{r}y \quad \text{iff} \quad x \succeq y.$$

If **r**′ *is any other such vector, then there exist* $\alpha > 0$ *and a vector* **b** *such that* $\mathbf{r}' = \alpha\mathbf{r} + \mathbf{b}$.

Proof: Let $\{\mathbf{x}_1, ..., \mathbf{x}_k\}$ $(k \leq n)$ be a set of nonnull, linearly independent vectors of labor inputs spanning the cone $\mathcal{L}$, chosen in such a way that $\mathbf{p}\hat{\mathbf{x}}_1 = \cdots = \mathbf{p}\hat{\mathbf{x}}_k$, where $\tilde{\mathbf{x}}_1 = [\mathbf{x}_1, \underline{\mathbf{x}}_1, \overline{\mathbf{x}}_1], \ldots,$ $\tilde{\mathbf{x}}_k = [\mathbf{x}_k, \underline{\mathbf{x}}_k, \overline{\mathbf{x}}_k]$ are production processes in P, in which $\mathbf{x}_1, \ldots,$ $\mathbf{x}_k$ appear as labor input vectors. By (P3), it follows that $\mathbf{x}_i \geq 0$ and so (since **p** is valid) $p = \mathbf{p}\hat{\mathbf{x}}_i > 0$ for all i $(i = 1, \ldots, k)$.

Let **X** be the matrix

$$\mathbf{X} = \begin{bmatrix} x_{11} & x_{21} & \cdots & x_{n1} & -p \\ x_{12} & x_{22} & \cdots & x_{n2} & -p \\ \vdots & \vdots & \ddots & \vdots & \vdots \\ x_{1k} & x_{2k} & \cdots & x_{nk} & -p \end{bmatrix}.$$

If the system $\mathbf{X}^T\mathbf{c} \geq 0$ has no solution, then Stiemke's Theorem[16] implies the existence of a vector $\mathbf{d} = [\delta_1 \cdots \delta_{n+1}] > 0$ such that $\mathbf{X}\mathbf{d}^T = 0$, i.e. such that

$$\delta_1 x_{11} + \quad \cdots \quad + \delta_n x_{n1} - \delta_{n+1} p \quad = 0$$

$$\vdots$$

$$\delta_1 x_{1k} + \quad \cdots \quad + \delta_n x_{nk} - \delta_{n+1} p \quad = 0$$

In such a case, setting $r_i = \delta_i / \delta_{n+1}$ and $\mathbf{r} = [r_1 \cdots r_n]$ we get

$$r_1 x_{11} + \quad \cdots \quad + r_n x_{n1} \quad = p$$

$$\vdots$$

$$r_1 x_{1k} + \quad \cdots \quad + r_n x_{nk} \quad = p$$

or, in abbreviated notation,

$$\mathbf{r} \mathbf{x}_i = p \qquad (i = 1, ..., k)$$

Thus, if $\mathbf{x}, \mathbf{y}$ are any elements of $\mathcal{L}$, then there are unique non-negative numbers $\alpha_1, ..., \alpha_k$ and $\beta_1, ..., \beta_k$, such that $\mathbf{x} = \sum_{i=1}^{k} \alpha_i \mathbf{x}_i$ and $\mathbf{y} = \sum_{i=1}^{k} \beta_i \mathbf{x}_i$. It follows that there are productive processes $\tilde{\mathbf{x}}$ and $\tilde{\mathbf{y}}$ in P such that $\tilde{\mathbf{x}} = [\mathbf{x}, \underline{\mathbf{x}}, \overline{\mathbf{x}}]$, $\tilde{\mathbf{y}} = [\mathbf{y}, \underline{\mathbf{y}}, \overline{\mathbf{y}}]$, $\hat{\mathbf{x}} = \overline{\mathbf{x}} - \underline{\mathbf{x}}$ and $\hat{\mathbf{y}} = \overline{\mathbf{y}} - \underline{\mathbf{y}}$, where $\underline{\mathbf{x}} = \sum_{i=1}^{k} \alpha \underline{\mathbf{x}}_i$, $\underline{\mathbf{y}} = \sum_{i=1}^{k} \beta \underline{\mathbf{x}}_i$, $\overline{\mathbf{x}} = \sum_{i=1}^{k} \alpha \overline{\mathbf{x}}_i$, $\overline{\mathbf{y}} = \sum_{i=1}^{k} \beta \overline{\mathbf{x}}_i$, $\hat{\mathbf{x}} = \sum_{i=1}^{k} \alpha \hat{\mathbf{x}}_i$, $\hat{\mathbf{y}} = \sum_{i=1}^{k} \beta \hat{\mathbf{x}}_i$. So we have

$$\mathbf{r}\mathbf{x} \geq \mathbf{r}\mathbf{y} \Leftrightarrow \sum_{i=1}^{k} \alpha_i \mathbf{r} \mathbf{x}_i \geq \sum_{i=1}^{k} \beta_i \mathbf{r} \mathbf{x}_i$$

$$\Leftrightarrow \left(\sum_{i=1}^{k} \alpha_i \right) p \geq \left(\sum_{i=1}^{k} \beta_i \right) p$$

$$\Leftrightarrow \mathbf{p}\hat{\mathbf{x}} \geq \mathbf{p}\hat{\mathbf{y}}$$

$$\Leftrightarrow \mathbf{x} \succsim \mathbf{y}.$$

Thus, it will suffice to show that $\mathbf{X}^T \mathbf{c} \geq \mathbf{0}$ has no solution. In order to derive a contradiction, assume that there is a solution $\mathbf{c} = [\gamma_1 \cdots \gamma_k]^T$ of $\mathbf{X}^T \mathbf{c} \geq \mathbf{0}$. Since $[-p \cdots -p] \mathbf{c} \geq 0$, it follows that not all the nonzero coordinates of $\mathbf{c}$ are positive. Analogously, since the remaining rows of $\mathbf{X}^T$ are semipositive, not all nonzero coordinates of $\mathbf{c}$ are negative. After column permutations and renumbering, we can let $\gamma_1, ..., \gamma_h$ be the nonnegative components and $\gamma_{h+1}, ..., \gamma_k$ be the negative ones of **c**. It follows that there are elements **u** and **v** in $\mathcal{L}$ which can be written as $\mathbf{u} = \sum_{i=1}^{h} \gamma_i \mathbf{x}_i$ and $\mathbf{v} = \sum_{i=h+1}^{k} (-\gamma_i) \mathbf{x}_i$. Since $[-p \cdots - p]\mathbf{c} = -p[\sum_{i=1}^{h} \gamma_i + \sum_{i=h+1}^{k} \gamma_i] \geq 0$, it follows that

(1)
$$\left(\sum_{i=h+1}^{k} (-\gamma_i) \right) p \geq \left(\sum_{i=1}^{h} \gamma_i \right) p$$

or, what is the same,

(2)
$$p\widehat{\mathbf{v}} \geq p\widehat{\mathbf{u}}.$$

Also, since the column vector that results from dropping the last component of $\mathbf{X}^T\mathbf{c}$ is nonnegative, and this is none other than the transpose of $\sum_{i=1}^{k} \gamma_i \mathbf{x}_i$, it follows that $\mathbf{u} = \sum_{i=1}^{h} \gamma_i \mathbf{x}_i \geq \sum_{i=h+1}^{k} (-\gamma_i) \mathbf{x}_i = \mathbf{v}$ and so, since **p** is admissible,

(3)
$$p\widehat{\mathbf{u}} \geq p\widehat{\mathbf{v}}.$$

Hence, from (2) and (3),

$$p\widehat{\mathbf{u}} = p\widehat{\mathbf{v}}.$$

On the other hand, the following chain of implications holds:

$$p\widehat{\mathbf{u}} = p\widehat{\mathbf{v}} \Rightarrow p(\widehat{\mathbf{u}} - \widehat{\mathbf{v}}) = 0$$
$$\Rightarrow p\left(\sum_{i=1}^{k} \gamma_i \widehat{\mathbf{x}}_i \right) = 0$$
$$\Rightarrow \sum_{i=1}^{k} \gamma_i p\widehat{\mathbf{x}}_i = 0$$
$$\Rightarrow \sum_{i=1}^{k} \gamma_i p = 0$$
$$\Rightarrow \sum_{i=1}^{k} \gamma_i (-p) = 0.$$

Thus, given that $\mathbf{X}^T\mathbf{c}$ is semipositive, it follows that $\mathbf{u} \geq \mathbf{v}$ and so that $\mathbf{p}\hat{\mathbf{u}} > \mathbf{p}\hat{\mathbf{v}}$. This contradiction shows that the system $\mathbf{X}^T\mathbf{c} \geq \mathbf{0}$ has no solution.

Any other reduction $\mathbf{r}'$ is a solution to

$$\mathbf{X}'(\mathbf{r}')^T = \begin{bmatrix} x_{11} & x_{21} & \cdots & x_{n1} \\ x_{12} & x_{22} & \cdots & x_{n2} \\ \vdots & \vdots & \ddots & \vdots \\ x_{1k} & x_{2k} & \cdots & x_{nk} \end{bmatrix} (\mathbf{r}')^T = \alpha \begin{bmatrix} p \\ \vdots \\ p \end{bmatrix}$$

for $\alpha > 0$. In particular, $\alpha\mathbf{r}^T$ is a solution to this same equation, since we have

$$\mathbf{X}'(\alpha\mathbf{r}^T) = \alpha \begin{bmatrix} p \\ \vdots \\ p \end{bmatrix}$$

Therefore, $(\mathbf{r}')^T$ differs from $\alpha\mathbf{r}^T$ by a vector $\mathbf{b}^T$ in the null space of $\mathbf{X}$, i.e. $\mathbf{r}' = \alpha\mathbf{r} + \mathbf{b}$. This establishes that reductions are unique up to affine transformations. $\square$

As I said before, if $\mathbf{r}$ is a reduction and $\mathbf{x} \in \mathcal{L}$, then the magnitude $\mathbf{rx}$ is called the labor-value of net output $\hat{\mathbf{x}}$. This concept is so important (in fact, this is the central concept of MTV) that it is worthy of being given a separate official definition.

DEFINITION 8: Let $\mathfrak{P} = \langle P_0, P \rangle$ be a productive structure, $\mathbf{p}$ a price system, and $\mathcal{L}$ an abstract labor structure corresponding to $\mathfrak{P}$ (i.e. to P) and $\mathbf{p}$. For any net output $\hat{\mathbf{x}}$ in the cone $\mathcal{C} = \{\hat{\mathbf{x}} \mid [\mathbf{x}, \underline{\mathbf{x}}, \overline{\mathbf{x}}] \in P$ and $\hat{\mathbf{x}} = \overline{\mathbf{x}} - \underline{\mathbf{x}}\}$, we define the *labor-value* of $\hat{\mathbf{x}}$, $\lambda(\hat{\mathbf{x}})$, as the number $\mathbf{rx}$, where $\mathbf{r}$ is a positive reduction.

Unlike the definition of labor-value provided by Marx in the prototype of MTV, that lead to a deviation of equilibrium prices and values, the definition of labor-value in terms of reductions leads directly to the following strong form of the Law of Value.

THEOREM 8: (Law of Value). *Let $\mathfrak{P}$ be a productive structure and $\mathbf{p}$ an equilibrium price for $\mathfrak{P}$, i.e. the profit rate of all nonnull processes in $\mathfrak{P}$ is uniform at $\mathbf{p}$. Then the labor-value of any net output vector $\hat{\mathbf{x}} \in \mathcal{C}$ is proportional to its price.*

Proof: It is immediate by Theorem 4 and the definition of $\mathfrak{P}$ that **p** is valid for $\mathfrak{P}$, and so there is an abstract labor structure $\langle \mathcal{L}, \succsim \rangle$ corresponding to $\mathfrak{P}$. Let **r** be any reduction representing this abstract labor structure such that $\mathbf{rx} = \mathbf{p}\hat{\mathbf{x}}$ for every $\tilde{\mathbf{x}} \in P$. If **r**′ is any other reduction, for fixed nonnull $\mathbf{x} \in \mathcal{L}$ let α be such that

$$\mathbf{r'x} = \alpha \mathbf{rx}.$$

Clearly, $\alpha > 0$. Let **y** be an arbitrary element of $\mathcal{L}$ and β the number such that $\mathbf{ry} = \beta \mathbf{rx} = \mathbf{r}(\beta \mathbf{x})$. Then $\mathbf{y} \sim \beta \mathbf{x}$ and so, since by hypothesis **r**′ is also a representation of $\succsim$, $\mathbf{r'y} = \mathbf{r'}(\beta \mathbf{x}) = \beta \mathbf{r'x} = \beta \alpha \mathbf{rx} = \alpha \mathbf{r}(\beta \mathbf{x}) = \alpha \mathbf{ry}$.

This chain of identities shows that the number α does not depend upon the particular vector of labor inputs **x**. $\square$

The Law of Value can be interpreted as establishing the reflection of all vectors of labor inputs into a particular one. For given reductions **r** and **r**′, we know by the Law of Value that there is a positive scalar α such that $\mathbf{r'x} = \alpha \mathbf{rx}$ for any $\mathbf{x} \in \mathcal{L}$. By the continuity of **r** (seen as a function from the cone $\mathcal{L}$ into $\mathbf{R}^+$), there is a vector of labor inputs $\mathbf{x}_0$ such that $\mathbf{rx}_0 = 1/\alpha$. For every $\mathbf{x} \in \mathcal{L}$,

$$\mathbf{rx} = (\mathbf{rx}/\mathbf{rx}_0)\mathbf{rx}_0 = \mathbf{r}((\mathbf{rx}/\mathbf{rx}_0)\mathbf{x}_0).$$

Hence,

$$\mathbf{x} \sim (\mathbf{rx}/\mathbf{rx}_0)\mathbf{x}_0 = (\alpha \mathbf{rx})\mathbf{x}_0 = (\mathbf{r'x})\mathbf{x}_0;$$

that is to say, the amount of abstract labor represented by **x** is reflected in $\mathbf{x}_0$ through the coefficient $\mathbf{r'x}$: vector **x** is equivalent to $\mathbf{r'x}$ times vector $\mathbf{x}_0$. Hence, at bottom, a change of reduction is just a change in the vector of labor inputs which is taken as unit. This should be compared with the change of unit rod in length measurement. From now on, whenever we talk of abstract labor, reductions, and labor-values, we shall suppose that in each case a particular vector of labor inputs has been chosen as unit of measurement.

Chapter 6

THE GENERAL AXIOMS OF THE THEORY

Along chapter 2 we saw the tribulations of the prototype of MTV in the face of the hard fact that the competition of capitals leads to a uniform profit rate, and so that values deviate from prices —contrary to what the Law of Value claims— unless we assume that the organic composition of capital is the same in all industries. Hence, it is easily seen that the notion of competition of capitals plays a central role in Marxian economics. Now, this competition of capitals requires to be effective a financial capital market, and therefore we are compelled to introduce a conceptual apparatus to describe such a market.

Also, the very notion of a uniform profit rate is introducing a concept of equilibrium within MTV. According to Marx, the aim of the capitalists as such is to maximize their benefits. As the classical economists would put it, the capitalists are looking after their own enlightened self-interest. In real life, however, the capitalists' looking after their own self-interest can adopt multiple forms. It can mean that the capitalists restrict their benefits for some time, to prevent a social problem from exploding, or that they are prepared to give money to other enterprises in order to prevent the collapse of the economic system. In other words, the capitalists can and do sometimes also act as *politicians* pursuing the welfare of their society, as a condition for the maintenance of institutions necessary to the functioning of the market. Indeed, it would be considered as stupid nowadays to restrict the role of the businessman to that of the brutal *homo œconomicus*, in disregard of all the other tasks he has to perform in his society and his responsibilities thereto.

As Hegel pointed out, there is more to the State than the mere protection of property. Yet, Hegel recognized that there is a sphere —a "moment"— in social life where the rules of the market have the upper hand and completely impose themselves in all activity or transaction. This is the moment of civil or burgeois society (*bürgerliche Gesellschaft*), the moment of "universal egoism". Clearly, MTV intends to discover and formulate the laws that govern this sphere. But, *within this sphere*, the enlightened self-interest of the capitalist is not to look after the general welfare as a citizen of the *polis*, but to look after the maximum benefit for his own enterprise. This is the kind of behavior with which MTV is concerned. Evidently, this behavior has been modeled by neoclassical economics through the notion of maximization of certain functions over production possibility sets that satisfy special conditions. This is of course an idealized depiction of the actual way in which entrepeneurs take decisions (usually under conditions of uncertainty), but a depiction that is necessary in order to logically derive the central results of MTV. In this respect of making strong assumptions to derive important results, MTV and the theory of competitive equilibrium do not differ at all.

Furthermore, it is plain that neoclassical equilibrium and MTV are complementary and conceptually akin —despite the long war between "bourgeois" and Marxian economists. We shall see that the Marxian notions of equilibrium and reproducibility naturally lead to the neoclassical concept of a competitive equilibrium. Hence, it is not to be fashionable that these notions are introduced here, but as a response to the very conceptual demands of MTV. This shall be apparent in the sequel.

In the first section I will provide a description of the kind of economic objects described by the theory; these turn out to be

Meinongian economies (in the sense of chapter 4), i.e. idealized nonexistent objects that bear some resemblance to modern market economies. Due to this reason, and to the fact that they are inspired in the description of the capitalist mode of production given by Marx, I will call them 'Marxian capitalist economies'. These economies are more general than the prototype of MTV as described in chapter 1, and comprise that model as a particular case. The fundamental law describing these economies is the Law of Profit Maximization, but the Law of Value is a logical consequence of the same. In other words, I shall prove the Law of Value out of the postulated behavior of the capitalists. This is given as a theorem in the same section. In the second section I present the proof of the most general version of the Fundamental Marxian Theorem, according to which the exploitation of the working class is necessary and sufficient for capitalist profit. In subsequent chapters the consistency and relevance of the general theory presented here will be apparent, where the possibility of the reproduction of the economy is proven, and an interesting specific model of the theory is developed in a quite detailed way.

6.1 A GENERAL MARKET ECONOMY

After the impressive collapse of the Soviet socialist monster, and the yearning of the Soviets and Eastern Europeans after market economies, the study of the essential nature of these economies has become more interesting than ever. What are the laws and conditions that make modern capitalist economies work so efficiently? In order to study in general the (reformed) Marxian response to this question, we shall consider a Meinongian economy which can be thought of as the capitalist economy in its ideal purity. This economy is constituted by l firms, represented by the numbers $1, \ldots, l$, collected in a set F. An arbitrary member of this set shall be denoted by the index f. Every firm $f \in F$ is governed by a decision organ —a president, a board of directors, or what have you— that has to face at the beginning of every economic cycle (say, a quarter or a year) a price system $\mathbf{p}$ in the standard simplex $S = \{[p_1 \cdots p_m] \mid \sum_{i=1}^{m} p_i = 1\}$ (the firms are price-takers) and decide on the basis of such price system which production process to operate among a set Y_f of possible production processes. The firm f possesses at the beginning of the cycle certain initial holdings devoted to production (it may have "in stock" other commodities) represented by a vector $\mathbf{h}_f \in \mathbb{R}^m$. These holdings are just enough for f to operate some process in Y_f, i.e. there is a production process $\tilde{\mathbf{x}} \in Y_f$ such that $\mathbf{h}_f = \underline{\mathbf{x}} + \mathbf{Bx}$. Firm f may operate this process, but it also has the option of obtaining or providing some credit in order to operate a different production process, including $\tilde{\mathbf{0}}$ (the firm has the possibility of inaction). I shall assume that the firm might be willing to transfer to other firms its whole initial vector of productive holdings, as well as to hire a credit with which it can buy all the initial holdings of the other firms. Hence, the set of credits possible for or by f will range from $-\mathbf{ph}_f$ to $\mathbf{ph} - \mathbf{ph}_f$. Notice that the range of credits available for any f depends on the prevalent system of prices and may vary as the price system changes. At any rate, I shall assume that the total social resources $\mathbf{t}$ (which must be greater than the aggregated initial productive holdings of the firms) are sufficient to operate any aggregated process $\tilde{\mathbf{x}} \in Y$ such that $\mathbf{px} + \mathbf{pBx} \leq \mathbf{ph}$, for any price system $\mathbf{p} \in S$, even though it might well be the case that $\underline{\mathbf{x}} + \mathbf{Bx} \geq \mathbf{h}$. In this last case, for example, the system of credit may be used to bring bundles not belonging to the sphere of production (the original $\mathbf{h}_f$'s), but rather to the stocks of some firm, in order to enlarge a bit the original global vector of initial endowments $\mathbf{h} = \sum_{f \in F} \mathbf{h}_f$ (this may happen when some prices in the system are zero, i.e. when some goods are free).

Given a price system $\mathbf{p}$ and before deciding which process to operate, firm f must ponder which production processes can operate using its initial resources and shop around a little to see if it can make some transactions allowing it to operate a different technology. In other words, the directives of any firm have to make up their minds as to which production process $\tilde{\mathbf{x}} \in Y_f$ are

they going to operate, so that —if they decide to do so— they
will be able to use a part or all of their initial holdings and ar-
range some credit in order to acquire (or give away) means of
production in the amounts c_f so that $h_f + c_f \leq \underline{x} + Bx$ for some
$\tilde{x} \in Y_f$. I shall assume that no matter how this decision is taken
(we shall see that profit maximizing drives the firms to use up
their resources to the maximum), the firm will always choose a
process in Y_f that maximizes profits at p subject to the constraint
that $px + pBx \leq ph_f + pc_f$. I will assume that the transferences
c_f of resources are paid at the end of the cycle (when all processes end simultaneously), and so the quantity pc_f is the credit
hired by the firm.

According to what has been said, the set of all possible transferences for firm f at price system p will be defined as

$$C_f(p) = \{c \,|\, h_f + c = \underline{x} + Bx \text{ and } p\underline{x} + pBx \leq ph\}$$

for some $\tilde{x} \in Y_f$. Since the monetary costs of these transferences
depend upon which price system is prevalent, the range of credits possible for f at p is the set

$$I_f(p) = \{pc \,|\, c \in C_f\}.$$

The financial capital market is depicted in the present theory
under the guise of credits available to or by the firms (the inter-
vals $I_f(p)$) and a uniform interest rate r at which credit is hired,
i.e. if firm f is lent (lends) c guilders, then it has to pay back (or
receive) $(1 + r)c$ guilders. As it can be shown, due to the conic
structure of the production possibilities sets it is really indiffer-
ent to the firms which credit they hire: their level of profits *in
production* is not affected by the amount of the credit taken or
given. Hence, profit maximization does not constrain the firms
to choose this or that credit; this is left to other considerations.
What matters here is that credits are available anyway, at a fixed
interest rate, and that it is possible for any financial capital to
obtain the same profit anywhere, even choosing the null process

(inaction), provided that they transfer the production means to be used up in some production process by another firm.

The production possibilities sets Y_f are far more general than the Leontief technology of the simple Marxian economy studied in chapter 1, §2. I suppose that such sets contain the possibility of inaction, i.e. the null process $\tilde{0}$. The technologies they represent yield constant returns to scale and so they are cones, but not necessarily polyhedric and not necessarily the same for all firms (there may be imperfect entry); the assumption of constant returns to scale implies, in particular, that all the commodities and amounts of labor are infinitely divisible. I suppose as well that the limit of every convergent sequence of production processes in any production set is also in the set, i.e. the sets are closed. Furthermore, I suppose that labor is indispensable to move the means of production ($\underline{\mathbf{x}} \geq \underline{\mathbf{0}} \Rightarrow \mathbf{x} \geq 0$), that there is no free lunch ($\overline{\mathbf{x}} \geq \overline{\mathbf{0}} \Rightarrow \mathbf{x} \geq 0$), and that labor is productive ($\overline{\mathbf{x}} \geq \overline{\mathbf{0}} \Leftarrow \mathbf{x} \geq 0$). The last assumption over the production possibilities sets is that the global set of possibilities of production, the aggregate set $Y = \sum_{h=1}^{l} Y_f$, is normal. What this means is that labor in all the available possible production processes is normal in the sense of Definition 6 of chapter 5. This assumption means that the labor firms can hire is immediately social, i.e. labor that has the average dexterity and skill, and that in all processes is put to work with an average speed and intensity.

As a result of the decision of the firms, a finite set P_0 of production processes is implemented. I shall consider the polyhedral cone P spanned by this set. This set P can be thought of as the technology actually chosen by the firms. A member of P_0 (one of the processes chosen by the firms) thus appears just as the technology P operated at a certain level. The decisions taken by the firms shall be represented by means of a function $d: F \rightarrow Y$ from the firms to the aggregate technology set Y, such that $d(h) \in Y$ is the decision taken by firm $f \in F$

There is a demand function for every kind of concrete labor. The "reproduction" of one hour of labor of type i ($i = 1, \ldots, n$) requires the consumption of the (column) vector $\mathbf{b}_i$ by the worker's household. All these consumption bundles are put together as columns of the matrix **B**, the consumption bundle of trade i being the ith column of **B**. We shall assume that the consumption bundles $\mathbf{b}_i$ belong to the cone $C = \{\hat{\mathbf{x}} \mid [\mathbf{x}, \underline{\mathbf{x}}, \overline{\mathbf{x}}] \in P$ and $\hat{\mathbf{x}} = \overline{\mathbf{x}} - \underline{\mathbf{x}}\}$ of net outputs of P.

These concepts are very abstract determinations. From a logico-mathematical point of view, they are abstract entities with certain mathematical properties; from an economic point of view they are notions that intend to describe historic categories corresponding to the capitalist mode of production. Taking these abstract concepts as a point of departure, I shall follow the dialectical method as it was explained on chapter 4 and build a class of structures, namely the class of Marxian capitalist economies, that intend to represent a particular class of states of a capitalist economy. Before I reach that concept, I would like to gather together in a unique axiomatic definition the former concepts, even though the mathematical characterization of some of them has been already given in previous chapters. To this effect I will introduce the concept of a basic Marxian structure.

DEFINITION 1: A *basic Marxian structure* is a structure of the form

$$\mathfrak{M} = \langle F, Y_f, S, \mathbf{h}_f, \mathbf{t}, d, r, \mathbf{B} \rangle$$

that satisfies the following axioms:

(1) There is a finite set of firms, represented by the set F of the first l positive integers.

(2) For every firm $f \in F$, the production possibility set Y_f of firm f is a subset of the nonnegative orthant Ω of $\mathbf{R}^{2m+n}$. Y_f is a closed convex cone in which labor is both productive and indispensable, and there is no free lunch. Furthermore, labor is normal in all production process available to the firms, i.e. the aggregate set Y is normal.

(3) The set of price systems is the standard simplex $S = \{[p_1 \cdots p_m] \mid \sum_{i=1}^{m} p_i = 1\}$

(4) The initial endowment $\mathbf{h}_f$ of every firm $f \in F$ is a semi-positive vector in $\mathbf{R}^m$ such that $\mathbf{h}_f = \underline{\mathbf{x}} + \mathbf{Bx}$ for some $\tilde{\mathbf{x}} \in Y_f$.

(5) The global social resources is a vector $\mathbf{t} \in \mathbf{R}^m$ such that $\mathbf{h} \leq \mathbf{t}$, where $\mathbf{h} = \sum_{h \in F} \mathbf{h}_f > \mathbf{0}$ is the aggregated vector of initial holdings of the firms. Moreover, for every $\mathbf{p} \in S$, $\underline{\mathbf{x}} + \mathbf{Bx} \leqq \mathbf{t}$ for every $\tilde{\mathbf{x}} \in Y$ such that $\mathbf{p}\underline{\mathbf{x}} + \mathbf{pBx} \leq \mathbf{ph}$.

(6) The decision function d is a function from F into the aggregate production set Y.

(7) The rate of interest r is a nonnegative real number.

(8) The demand matrix $\mathbf{B}$ is a $m \times n$ matrix. Every column $\mathbf{b}_i$ of this matrix is semipositive and is in $\mathcal{C}$ for every $i = 1, \ldots, n$ and, for every kind of good $i = 1, \ldots, m$: either there is a process $\tilde{\mathbf{x}} \in Y$ such that $\bar{x}_j = b$, where b is a positive entry in the matrix $\mathbf{B}$ of consumption bundles for the working class, or there is a process $\tilde{\mathbf{x}} \in Y$ such that $\underline{x}_i > 0$, but not both.

Definition 1 provides the basic conceptual framework, together with the mathematical characterization of the concepts of the theory, which is required to formulate the substantial axioms of MTV. Notice that Axiom (8) implies that there are two mutually exclusive kinds of goods in the economy: wage and capital goods.

Before the substantial axioms of MTV can be introduced, it is necessary to develop a little bit this conceptual apparatus. This is the aim of the following definitions and statements

DEFINITION 2: Let $\mathbf{p}$ be a given price system and define

(1) The *set of all possible transferences for firm* f at $\mathbf{p} \in S$ is

$$C_f(\mathbf{p}) = \{\mathbf{c} \mid \mathbf{h}_f + \mathbf{c} = \underline{\mathbf{x}} + \mathbf{Bx}, \ \tilde{\mathbf{x}} \in Y_f, \ \mathbf{p}\underline{\mathbf{x}} + \mathbf{pBx} \leq \mathbf{ph}\}$$

(2) The *set of credits possible for firm* f *at* $\mathbf{p} \in S$ is

$$I_f(\mathbf{p}) = \{\mathbf{pc} \mid \mathbf{c} \in C_f\}.$$

(3) The *set of all possible initial monetary resources for firm f* at price system $\mathbf{p} \in S$ is

$$\Xi_f(\mathbf{p}) = \{\mathbf{ph}_f + c_f \,|\, c_f \in I_f(\mathbf{p})\}$$

I shall prove now that the just defined objects have certain properties which are important for the development of the theory. Recall that the norm $\|\mathbf{v}\|$ of a vector $\mathbf{v} = [v_1 \cdots v_s]$ is its distance to the origin of the linear space to which it belongs, given by the formula $\|\mathbf{v}\| = \sqrt{v_1^2 + \cdots + v_s^2}$. Say that a set $X \subseteq Y$ of production processes is *bounded* iff the set of the norms of all processes in X is bounded, i.e. the distance of any of these processes to the null process (the origin) is never greater than a specified positive real number.

LEMMA 1: *If $\bar{X}$ is a subset of Y such that $X = \{\mathbf{x} \,|\, [\mathbf{x}, \underline{\mathbf{x}}, \bar{\mathbf{x}}] \in \bar{X}\}$ is bounded, then $\bar{X}$ is also bounded.*

Proof: If $\bar{X}$ were unbounded, there would be an unbounded sequence in $\bar{X}$. Thus, it will be sufficient to show that every sequence in $\bar{X}$ is bounded. To this effect, assume that $(\tilde{\mathbf{x}}_k)$ is any sequence in $\bar{X}$. If $(\tilde{\mathbf{x}}_k)$ were unbounded, it would have a subsequence —call it also $(\tilde{\mathbf{x}}_k)$— such that $(\|\tilde{\mathbf{x}}_k\|)$ would be increasing and unbounded. At any rate, the corresponding sequence of labor input vectors $(\mathbf{x}_k)$ can be assumed to converge to a limit $\mathbf{x}$ (not necessarily in X) because X is bounded. Let

$$\tilde{\mathbf{y}}_k = (\|\tilde{\mathbf{x}}_k\| + 1)^{-1}\tilde{\mathbf{x}}_k.$$

Since Y is a cone, $\tilde{\mathbf{y}}_k \in Y$. Moreover, $(\tilde{\mathbf{y}}_k)$ is bounded because $\|\tilde{\mathbf{y}}_k\| \leq 1$. Hence, without loss of generality we may assert that $(\tilde{\mathbf{y}}_k)$ converges to a point $\tilde{\mathbf{y}}$ which must belong to Y, because Y is closed. Since $(\|\tilde{\mathbf{y}}_k\|)$ is increasing, $\tilde{\mathbf{y}} \neq \tilde{\mathbf{0}}$ and so, by Axiom 2 of Definition 1, $\bar{\mathbf{y}} \geq \mathbf{0}$. On the other hand, since $(\mathbf{x}_k) \to \mathbf{x}$,

$$\lim_{k \to \infty} \mathbf{y}_k = \lim_{k \to \infty} (\|\tilde{\mathbf{x}}_k\| + 1)^{-1} \cdot \lim_{k \to \infty} \mathbf{x}_k$$
$$= 0 \cdot \mathbf{x}$$
$$= 0.$$

Hence, be the same axiom again, $\tilde{\mathbf{y}} = \tilde{\mathbf{0}}$. This contradiction shows that $(\tilde{\mathbf{x}}_k)$ is bounded. $\square$

LEMMA 2:   *For every $f \in F$ and $\mathbf{p} \in S$, $I_f(\mathbf{p})$ is a nonempty compact interval of real numbers, and so it is $\varXi_f(\mathbf{p})$.*

Proof: I will show, succesively, that $I_f(\mathbf{p})$ is nonempty, bounded, connected and closed. Since $\varphi(\mathbf{c}) = \mathbf{pc}$ is continuous on C_f for every $\mathbf{p} \in S$, $I_f(\mathbf{p})$ is nothing but the range of φ, and convexity implies connectedness, it will be sufficient to show that C_f is nonempty, bounded, closed and convex.

First of all, since $\mathbf{h}_f + \mathbf{0} = \mathbf{h}_f = \underline{\mathbf{x}} + \mathbf{Bx}$ for some $\tilde{\mathbf{x}} \in Y_f$ and $\mathbf{ph}_f \le \mathbf{ph}$, it follows that $\mathbf{0} \in C_f(\mathbf{p})$.

Since the set $X = \{\mathbf{x} \,|\, \mathbf{pBx} \le \mathbf{ph}\}$ is bounded, by Lemma 1 the set $\tilde{X} = \{\tilde{\mathbf{x}} \,|\, \mathbf{pBx} \le \mathbf{ph}\}$ is also bounded. It follows that $\{\underline{\mathbf{x}} + \mathbf{Bx} - \mathbf{h}_f \,|\, \tilde{\mathbf{x}} \in \tilde{X}\} = C_f(\mathbf{p})$ is bounded.

Let $(\mathbf{c}_k)$ be a sequence in $C_f(\mathbf{p})$ converging to a vector $\mathbf{c}$. It follows that there is a sequence $(\underline{\mathbf{x}}_k + \mathbf{Bx}_k)$ in $\{\underline{\mathbf{y}} + \mathbf{By} \,|\, \tilde{\mathbf{y}} \in \tilde{X}\}$ such that $\underline{\mathbf{x}}_k + \mathbf{Bx}_k = \mathbf{h}_f + \mathbf{c}_k$ and $\mathbf{p}\underline{\mathbf{x}}_k + \mathbf{pBx}_k \le \mathbf{ph}$. Let $(\tilde{\mathbf{x}}_k)$ be the sequence of production processes in $\tilde{X}$ corresponding to the inputs $(\underline{\mathbf{x}}_k + \mathbf{Bx}_k)$. Since $\tilde{X}$ is bounded, without loss of generality we may suppose that $(\tilde{\mathbf{x}}_k)$ converges to a point $\tilde{\mathbf{x}}$ which must be in Y_f, because Y_f is closed. It is easy to see that

$$\lim_{k \to \infty}(\underline{\mathbf{x}}_k + \mathbf{Bx}_k) = \underline{\mathbf{x}} + \mathbf{Bx}$$

and also that

$$\mathbf{p}\underline{\mathbf{x}} + \mathbf{pBx} \le \mathbf{ph}.$$

The verification that $C_f(\mathbf{p})$ is convex is simple and is left to the reader.

Since all the credits that f can give or take have the structure

$$c = \mathbf{p}\underline{\mathbf{x}} + \mathbf{pBx} - \mathbf{ph}_f$$

for $\tilde{\mathbf{x}} \in Y_f$ and $\mathbf{p}\underline{\mathbf{x}} + \mathbf{pBx} \le \mathbf{ph}$, the maximum point of $I_f(\mathbf{p})$ is $\mathbf{ph} - \mathbf{ph}_f$, whereas the minimum is $-\mathbf{ph}_f$. In other words, $I_f(\mathbf{p})$

is the interval $[-\mathbf{ph}_f, \mathbf{ph} - \mathbf{ph}_f]$. Hence, $\Xi_f(\mathbf{p})$ is nothing but the translated interval $[0, \mathbf{ph}]$. $\square$

We have shown, incidentally, that the set of financial resources $\Xi_f(\mathbf{p})$ depends on the price system but not on the firm f. Hence, from now on I will drop the subindex 'f' from all such expressions, except when it is necessary to stress that it is referred to some firm. When $\mathbf{p}$ is specified, the variable ξ will range over the elements of $\Xi(\mathbf{p})$. Hence, it can be substituted by an expression of the form '$\mathbf{ph}_f + c$', where c is a variable ranging over $I_f(\mathbf{p})$.

DEFINITION 3: Let a basic Marxian structure $\mathfrak{M}$ be given. For each firm $f \in F$ we have the following concepts:

(1) For each $\mathbf{p} \in S$, let $\Xi(\mathbf{p})$ be the interval $[0, \mathbf{ph}]$.

(2) The *financial feasibility function* for firm f is the mapping $\mathbf{B}_f: S \times \Xi \to Y_f$ such that, for every $(\mathbf{p}, \xi) \in S \times \Xi$,

$$B_f(\mathbf{p}, \xi) = \{\tilde{\mathbf{x}} \in Y_f \mid \mathbf{p}\underline{\mathbf{x}} + \mathbf{pBx} \leq \xi\}.$$

Sometimes I shall write $B_f(\mathbf{p}, c)$, when I need to point out that ξ is $\mathbf{ph}_f$ plus the credit c.

(3) The *profit maximizing function* for f is the mapping $\Pi_f: S \times \Xi \to \mathbf{R}$ defined by

$$\Pi_f(\mathbf{p}, \xi) = \max\{\mathbf{p}\hat{\mathbf{x}} - \mathbf{pBx} - rc \mid \tilde{\mathbf{x}} \in B_f(\mathbf{p}, \xi)\}.$$

where $\xi = \mathbf{ph}_f + c$ for $c \in I_f(\mathbf{p})$.

(4) The *optimality function* for f is the mapping $A_f: S \times \Xi \to Y_f$ such that

$$A_f(\mathbf{p}, \xi) = \{\tilde{\mathbf{x}} \in B_f(\mathbf{p}, \xi) \mid \mathbf{p}\hat{\mathbf{x}} - \mathbf{pBx} = \Pi_f(\mathbf{p}, \xi)\}.$$

As before, sometimes I shall write $A_f(\mathbf{p}, c)$, when I need to point out that ξ is $\mathbf{ph}$ plus the credit c.

(5) The *global financial feasibility function* is the mapping $\mathbf{B} : S \times \Xi \rightarrow Y$ such that, for every $(\mathbf{p}, \xi) \in S \times \Xi$,

$$B(\mathbf{p}, \xi) = \{\tilde{\mathbf{x}} \in Y \mid \mathbf{p}\underline{\mathbf{x}} + \mathbf{p}\mathbf{B}\mathbf{x} \leq \xi\}.$$

(6) The *global profit maximizing function* is the mapping $\Pi : S \times \Xi \rightarrow \mathbf{R}$ defined by

$$\Pi(\mathbf{p}, \xi) = \max\{\mathbf{p}\hat{\mathbf{x}} - \mathbf{p}\mathbf{B}\mathbf{x} \mid \tilde{\mathbf{x}} \in B(\mathbf{p}, \xi)\}.$$

(7) The *global optimality function* is the mapping $A : S \times \Xi \rightarrow Y$ such that

$$A(\mathbf{p}, \xi) = \{\tilde{\mathbf{x}} \in B(\mathbf{p}, \xi) \mid \mathbf{p}\hat{\mathbf{x}} - \mathbf{p}\mathbf{B}\mathbf{x} = \Pi(\mathbf{p}, \xi)\}.$$

(8) P_0, the set of all *actually operated processes* in $\mathfrak{M}$ is the set of all processes actually chosen by the firms:

$$\{d(h) \mid f \in F\}.$$

(9) P, the *actual technology* of $\mathfrak{M}$, is the convex polyhedral cone spanned by P_0.

(10) $\mathcal{L}$, the *set of labor inputs* of $\mathfrak{M}$, is the set of all vectors of labor inputs of the processes in the actual technology of $\mathfrak{M}$:

$$\mathcal{L} = \{\mathbf{x} \mid [\mathbf{x}, \underline{\mathbf{x}}, \overline{\mathbf{x}}] \in P\}.$$

(11) The price system $\mathbf{p}$ is said to be *feasible* for firm $f \in F$ iff $\Pi_f(\mathbf{p}, c) > 0$ for some $c \in I_f$. We say that a price system is *feasible* if it is feasible for every firm.

(12) A good of type i ($i = 1, \ldots, m$) is called a *wage good* if there is a process $\tilde{\mathbf{x}} \in Y$ such that $\overline{x}_j = b$, where b is a positive entry in the matrix $\mathbf{B}$ of consumption bundles for the working class. A good of type i ($i = 1, \ldots, m$) is called a *capital good* in the economy if there is a process $\tilde{\mathbf{x}} \in Y$ such that $\underline{x}_i > 0$.

Less precise definitions of concepts (2)-(4) are due originally to Roemer (1981).[1] It is illustrative to see the motivation behind the definition of Π_f. Actually, if credit c is borrowed, the profit is

$$\max\{\mathbf{p\bar{x}} + [c + \mathbf{ph}_f - (\mathbf{p\underline{x}} + \mathbf{pBx})]$$
$$-[(1 + r)c] - \mathbf{ph}_f \mid \tilde{\mathbf{x}} \in B_f(\mathbf{p}, c)\},$$

"where the terms are, respectively, income from production, the value of assets not used in production but held over to the next period, the value of borrowing repaid, and the value of today's endowments".[2] But the expressions within the brackets can be simplified to obtain clause (2) of the definition. Using the notions introduced in Definition 3 it is possible to formulate the fundamental law that characterizes the Marxian capitalist economies. This is the Law of Profit Maximization, a law about the behavior of the capitalist as such, according to which he always choses to operate a productive process that maximizes his profit. The Law of Value (which is the law that defines MTV, as I said in chapter 2, §1), as well as the Fundamental Marxian Theorem, can be derived from this axiom and so, in effect, MTV is a logical consequence of the theory determined by the following definition.

DEFINITION 4: A *Marxian capitalist economy* is a basic Marxian structure in which all firms maximize their profit with respect to a feasible system of financial resources. That is to say, the following law holds for every $f \in F$ and $c \in I_f$:

$$d(h) \in A_f(\mathbf{p}, \xi_f) \quad \text{and} \quad \sum_{f \in F} \xi_f = \mathbf{ph}.$$

Definition 4 defines the models of the Marxian economic theory. The Law of Value has to be proven to hold as a consequence of the Law of Profit Maximization. In Marxian dialectical terms this means that the "essence" is implied by the "phenomenon", but this terminology is not well applied here, because it is as essential to capitalism profit maximization as it is the reduction of concrete to abstract labor. In deriving the Law of Value from

profit maximization we in effect establish that the process of reduction of heterogeneous labors is effected by that behavior. This
is the import of the first theorem of the chapter, whose demon-
stration requires the support of two previous lemmas.

LEMMA 3: *In order to maximize its profits firm $f \in F$ activates any
chosen nonnull technology to the limit of its resources; i.e. if $\tilde{\mathbf{x}} \in A_f(\mathbf{p}, c)$
then* $\mathbf{p}\mathbf{x} + \mathbf{p}\mathbf{B}\mathbf{x} = \mathbf{p}\mathbf{h}_f + c$.

Proof: At every point $\tilde{\mathbf{x}}$ of $A_f(\mathbf{p}, c)$ the following equation is satisfied:

$$\mathbf{p}\underline{\mathbf{x}} + \mathbf{p}\mathbf{B}\mathbf{x} = \mathbf{p}\mathbf{h}_f + c.$$

In effect, if $\tilde{\mathbf{y}}$ is such that

$$\mathbf{p}\underline{\mathbf{y}} + \mathbf{p}\mathbf{B}\mathbf{y} < \mathbf{p}\mathbf{h}_f + c,$$

then there is an α such that

$$\alpha(\mathbf{p}\underline{\mathbf{y}} + \mathbf{p}\mathbf{B}\mathbf{y}) = \mathbf{p}\mathbf{h}_f + c.$$

Let $\tilde{\mathbf{z}}$ be the process $\alpha\tilde{\mathbf{y}}$. Then $\mathbf{p}\underline{\mathbf{z}} + \mathbf{p}\mathbf{B}\mathbf{z} = \mathbf{p}\mathbf{h}_f + c$ and, moreover,

$$\mathbf{p}\widehat{\mathbf{z}} - \mathbf{p}\mathbf{B}\mathbf{z} - rc = \alpha(\mathbf{p}\widehat{\mathbf{y}} - \mathbf{p}\mathbf{B}\mathbf{y}) - rc$$
$$> \mathbf{p}\widehat{\mathbf{y}} - \mathbf{p}\mathbf{B}\mathbf{y} - rc$$

because $\alpha > 1$. It is thus seen that at any rate it is more profitable
for f to activate the process $\tilde{\mathbf{z}} = \alpha\tilde{\mathbf{y}}$ than the less intense process
$\tilde{\mathbf{y}}$. $\square$

LEMMA 4: *If process $\tilde{\mathbf{x}}$ maximizes profits at credit c then, for every
credit $c' \in I_f(\mathbf{p})$, there exists a positive α such that $\alpha\tilde{\mathbf{x}}$ maximizes profits
at credit c'.*

Proof: Assume that $\tilde{\mathbf{x}} \in A_f(\mathbf{p}, c)$ and suppose that, for every $\alpha >
0$, $\alpha\tilde{\mathbf{x}} \notin A_f(\mathbf{p}, c')$. Then, in particular, $\alpha\tilde{\mathbf{x}} \notin A_f(\mathbf{p}, c')$ for $\mathbf{p}(\alpha\underline{\mathbf{x}}) +
\mathbf{p}\mathbf{B}(\alpha\mathbf{x}) = \mathbf{p}\mathbf{h}_f + c'$. Hence, there is[3] a $\tilde{\mathbf{y}} \in A_f(\mathbf{p}, c)$ such that
$\mathbf{p}\underline{\mathbf{y}} + \mathbf{p}\mathbf{B}\mathbf{y} = \mathbf{p}\mathbf{h}_f + c'$ and

$$\mathbf{p}\widehat{\mathbf{y}} - \mathbf{p}\mathbf{B}\mathbf{y} - rc' > \mathbf{p}(\alpha\widehat{\mathbf{x}}) - \mathbf{p}\mathbf{B}(\alpha\mathbf{x}) - rc'.$$

Thus, if β is such that $\beta(\mathbf{ph}_f + c') = \mathbf{ph}_f + c$, then

$$\beta\alpha(\mathbf{p}\underline{x} + \mathbf{pBx}) = \mathbf{ph}_f + c = \mathbf{p}\underline{x} + \mathbf{pBx}$$

and so $\beta = \alpha^{-1}$.

Therefore,

$$\alpha^{-1}(\mathbf{p}\underline{y} + \mathbf{pBy}) = \alpha^{-1}(\mathbf{ph}_f + c') = \mathbf{ph}_f + c$$

and yet

$$\mathbf{p}(\alpha^{-1}\widehat{\mathbf{y}}) - \mathbf{pB}(\alpha^{-1}\mathbf{y}) - rc > \mathbf{p}\widehat{\mathbf{x}} - \mathbf{pBx} - rc.$$

This implies that $\widetilde{\mathbf{x}} \in B_f(\mathbf{p}, c)$ and also $\widetilde{\mathbf{x}} \notin A_f(\mathbf{p}, c)$, a contradiction. $\square$

THEOREM 1: *The Law of Value holds true in a Marxian capitalist economy, provided that the price system is feasible.*

Proof: At point $c \in I_f$, Π_f adopts the value

$$\begin{aligned}
\Pi_f(\mathbf{p}, c) &= \mathbf{p}\widehat{\mathbf{x}} - \mathbf{pBx} - rc \\
&= \pi(\widetilde{\mathbf{x}}) \cdot (\mathbf{ph}_f + c) - rc
\end{aligned}$$

where $\pi(\widetilde{\mathbf{x}})$ is the rate of profit of $\widetilde{\mathbf{x}}$ as defined in Definition 4 of Chapter 5. Actually, by Lemma 4, if $\widetilde{\mathbf{y}}$ maximizes Π_f at c', then there is an α such that $\alpha\widetilde{\mathbf{x}}$ also maximizes Π_f at c' and so

$$\pi(\widetilde{\mathbf{x}}) = \pi(\alpha\widetilde{\mathbf{x}}) = \pi(\widetilde{\mathbf{y}}),$$

i.e. the profit rate π of the profit-maximizing processes is constant at all credits $c \in I_f$. Hence, we can write

$$\begin{aligned}
\Pi_f(\mathbf{p}, c) &= \pi(\mathbf{ph}_f + c) - rc \\
&= \pi\mathbf{ph}_f + \pi c - rc \\
&= \pi\mathbf{ph}_f + (\pi - r)c.
\end{aligned}$$

It is obvious that Π_f is differentiable with respect to c, its derivative being

$$\frac{\partial \Pi_f}{\partial c} = \pi - r.$$

Since I_f is compact, it follows that Π_f assumes a maximum at some point $c^* \in I_f$ and so, at that point,

$$\frac{\partial \Pi_f}{\partial c} = \pi - r = 0.$$

This shows that the interest rate is actually equal to the profit rate and that Π_f is constant, its form being exactly

$$\Pi_f(\mathbf{p}, c) = r\mathbf{ph}_f,$$

i.e. the function Π_f refers only to the profit obtained over the initial endowments, not over these plus the credits hired.

By the Law of Profit Maximization, all processes $\tilde{\mathbf{x}}_1, \ldots, \tilde{\mathbf{x}}_l$ in P_0, the processes chosen by the firms, have the same profit rate r, which is positive because the price system is feasible. Since the positive hull P^+ of P_0 is a subset of Y, and Y is normal (by Axiom 2 of Definition 1), it follows by Theorem 2 of chapter 5 that P^+ is also normal. Hence, by Theorem 3 of the same chapter, the profit rate of all processes in P^+ is uniform and equal to r. By Theorem 4, $\mathbf{p}$ is valid for P, the actual technology of $\mathfrak{M}$, and so it induces an abstract labor relation $\succeq$ over the set $\mathcal{L}$ of labor input vectors of such technology. In other words, $\langle \mathcal{L}, \succeq \rangle$ is the abstract labor structure corresponding to the price system $\mathbf{p}$ and the technology P (Definition 3, chapter 5). Since $\mathfrak{P} = \langle P_0, P \rangle$ is a productive structure, by Axiom 2 of Definition 1, Theorem 8 of chapter 5 yields the desired conclusion. $\square$

It is fashionable nowadays to speak of the "efficiency" of market economies. One way of making sense of this phrase is to say that the egoist behavior of the firms guarantees that the most efficient processes (in the sense of Definition 5, chapter 5) are operated, consistent with the restriction in capital stocks and credit. This is certainly true in a Marxian capitalist economy in which there are no free goods and is established here as a theorem.

THEOREM 2: *If the price system is positive and feasible, the process chosen by each firm is efficient.*

Proof: Let $\tilde{\mathbf{x}}$ be the process chosen by firm $f \in F$ and $\mathbf{w} = \mathbf{pB}$. If $\tilde{\mathbf{x}}$ were not efficient, there would be a process $\tilde{\mathbf{y}} \in Y_f$ such that $\mathbf{y} \leq \mathbf{x}$ and $\hat{\mathbf{y}} \geq \hat{\mathbf{x}}$, with at least one of these two inequalities being strict. Since $\mathbf{p}, \mathbf{w} > 0$, we have $\mathbf{wy} < \mathbf{wx}$ or $\mathbf{p}\hat{\mathbf{y}} > \mathbf{p}\hat{\mathbf{x}}$ and so, at any rate, $\mathbf{p}\hat{\mathbf{y}} - \mathbf{wy} > \mathbf{p}\hat{\mathbf{x}} - \mathbf{wx}$. Since Y is normal, $\mathbf{y} \leq \mathbf{x}$ implies $\underline{\mathbf{y}} \leq \underline{\mathbf{x}}$ and so $\mathbf{p}\underline{\mathbf{y}} + \mathbf{wy} \leq \mathbf{p}\underline{\mathbf{x}} + \mathbf{wx} \leq \mathbf{ph} + c$. This means that the profit at $\tilde{\mathbf{y}} \in B_f(\mathbf{p}, c)$ is greater than that at $\tilde{\mathbf{x}}$, contradicting the assumption. $\square$

Having established how the behavior of the capitalists implies the Law of Value, as well as the efficiency of market economies, I shall proceed now to prove the Fundamental Marxian Theorem, which is also a general theorem that follows from the essence of Marxian capitalist economies.

6.2 THE FUNDAMENTAL MARXIAN THEOREM

Perhaps the most classical and single important result of MTV is the proof that the exploitation of the workers is both a necessary and sufficient condition for the existence of profits in the capitalist firms. We already saw in chapter 1 this proof for the case of the protoype. In that chapter we saw Marx's definition of the exploitation rate, namely as the ratio of surplus-value to necessary labor:

$$\varepsilon = \frac{x - v}{v}$$

where x_i is total amount of live labor in the process operated at the unitary level and v is the "necessary labor" of the same, i.e. the value of the subsistence basket that the workers of the process are able to buy with their salary. Since we are no longer assuming that labor is homogeneous and that the technology is of Leontief type, we need to redefine these notions.

The total amount of live labor is associated now with the *net-output* of the process $\tilde{\mathbf{x}}$, and is none other than the labor-value of the net output of this process: $\lambda(\hat{x})$. The value of the subsistence basket of the workers is equal to the labor-value $\lambda(b)$ of the total consumption basket for the workers of the process, where that basket is given by $b = Bx$.

Hence, the exploitation rate at a non-null process $\tilde{x}$ is just

$$\varepsilon(\tilde{x}) = \frac{\lambda(\hat{x}) - \lambda(b)}{\lambda(b)}.$$

(By convention, we say that the exploitation rate of the null process $\tilde{0}$ is 0.)

One of the aims of Marx in *Capital* was to unveil the fact that the exploitation of the workers by the capitalists is the "source" of the latter's profits. A precise formulation of this claim is the Fundamental Marxian Theorem, that was proven for the first time for a Leontief technology with homogeneous labor by Morishima and Seton (1961), and Okishio (1963) (see chapter 1 for the details of the proof for the prototype of MTV). A proof of the theorem for a von Neumann economy with homogeneous labor is provided by Morishima (1974). A proof of the theorem for a Leontief technology with heterogeneous labor is given by Krause (1981). The following version is the most general thus far.

THEOREM 3: (The Fundamental Marxian Theorem). *For every labor process $\tilde{x} \in P$: the profit rate $\pi(\tilde{x})$ of $\tilde{x}$ is positive if, and only if, the rate of exploitation $\varepsilon(\tilde{x})$ of $\tilde{x}$ is positive.*

Proof: First of all, by the Law of Value, we may suppose that $p\hat{x} = \lambda(\tilde{x})$ for every $\tilde{x} \in P$. Let $\tilde{x}$ be any process in the actual technology P and suppose first that its profit rate $\pi(\tilde{x})$ is equal to the uniform prevalent profit rate π, which is positive. This means that

$$\frac{p\hat{x} - pBx}{p\underline{x} + pBx} > 0.$$

Hence, since $p\underline{x} + pBx$ cannot be negative, $p\hat{x} - pBx = p\hat{x} - pb = \lambda(\hat{x}) - \lambda(b) > 0$. Since $b = \hat{y}$ for some $\tilde{y} \in P$ and $\pi(\tilde{y}) = \pi > 0$, it follows that $\lambda(\mathbf{b}) = \mathbf{pb} > 0$ and therefore the number

$$\varepsilon(\widetilde{\mathbf{x}}) = \frac{\lambda(\widehat{\mathbf{x}}) - \lambda(\mathbf{b})}{\lambda(\mathbf{b})}$$

is defined and positive.

Conversely, if

$$\frac{\lambda(\widehat{\mathbf{x}}) - \lambda(\mathbf{b})}{\lambda(\mathbf{b})} > 0,$$

since $\lambda(\mathbf{b}) = \mathbf{pb} = \mathbf{pBx}$ cannot be negative, it must needs be positive, and so $\lambda(\widehat{\mathbf{x}}) - \lambda(\mathbf{b}) = \mathbf{p\widehat{x}} - \mathbf{pBx} > 0$. It follows that

$$\pi(\widetilde{\mathbf{x}}) = \frac{\mathbf{p\widehat{x}} - \mathbf{pBx}}{\mathbf{p\underline{x}} + \mathbf{pBx}} > 0. \quad \square$$

This theorem is absolutely general since it follows only from the assumptions that define the concept of a Marxian capitalist economy. More than a mere "denunciation" of capitalism, this theorem is a very important explanatory condition for the reproduction of market economies. Besides, the concept of exploitation should not be confused with the concept of alienation: there may be exploitation without alienation and viceversa. But nevertheless there is a tendency toward the increase of the exploitation rate within capitalism in the form of a struggle of the particular capitalist to keep the wages of his employees to the minimum.

A consequence of the FMT is that exploitation is necessary for the feasibility of price systems. In order for a price system to be feasible, it is necessary that at that price every firm must be able to operate a process at which the exploitation rate is positive. I shall conclude the present chapter proving this interesting result.

THEOREM 4: *If the price system is feasible then every firm operates a production process whose exploitation rate is positive.*

Proof: Suppose that **p** is a feasible price system. Then every firm can operate a nonnull process yielding positive profits, and so it does it by the Law of Profit Maximization. Hence, the profit rate of all processes operated by the firms is positive and therefore at price **p** the exploitation rate of all the processes chosen by the firms is positive. □

Chapter 7

GENERAL REPRODUCIBILITY

The aim of the present chapter is to explore the conditions of reproducibility of a general Marxian capitalist economy. I want to consider here the conditions that make it possible for a market economy to reproduce itself. Is it sufficient for reproducibility to let the firms just pursue their enlightened self-interest, or something more is required? Is it even possible to make compatible that behavior with reproducibility? These are the leading questions of this chapter. After introducing a formal definition of reproducibility, I will develop the conceptual apparatus to some extent, in order to prove the existence of what I shall call a 'reproducible global decision' (RGD), i.e. to prove that the firms need not but can choose a technology that maximizes their profits *and* also guarantees the reproducibility of the economy. This result shall be established as the existence of a certain competitive equilibrium. Since the concept of (simple or extended) reproducibility is quite tipically Marxian, it will be evident that the conceptual framework of general equilibrium theory and that of MTV are deeply related.

We have seen already (in the previous chapter) that the behavior of the firms guarantees the equalization of the profit rate, but there is more to equilibrium than just a uniform profit rate. The concept of RGD equilibrium we are interested in includes several features: (1) it is a selection of production processes by the firms such that the process chosen by each firm maximizes the benefit for that firm; (2) the processes are able to reproduce themselves, in the sense that the aggregate process produces at least what it consumes in terms of wage and capital goods, and perhaps some excedent; (3) the operation of the production processes is feasible, in the sense that there are sufficient stocks accumulated in the economy to implement the global process $\tilde{\mathbf{x}}$. In symbolic terms, these features can be expressed as follows.

DEFINITION 1: $\{\tilde{\mathbf{x}}_1, \ldots, \tilde{\mathbf{x}}_l\}$ is a *reproducible global decision* (RGD) iff there exist credits $c_1, \ldots, c_l$ and a price system $\mathbf{p}$ such that

  (1) For every firm $f \in F$, process $\tilde{\mathbf{x}}_f$ maximizes profits for f, i.e. $\tilde{\mathbf{x}}_f \in A_f(\mathbf{p}, c_f)$
  (2) The global process $[\mathbf{x}, \underline{\mathbf{x}}, \overline{\mathbf{x}}] = \sum_{f \in F} \tilde{\mathbf{x}}_f$ is nonnull and reproducible, i.e. $\overline{\mathbf{x}} \geq \underline{\mathbf{x}} + \mathbf{Bx}$
  (3) The global process is feasible, in the sense that there are sufficient stocks accumulated in the economy to implement the global process, i.e. $\underline{\mathbf{x}} + \mathbf{Bx} \leq \mathbf{t}$.
  (4) The capital market clears, i.e. $\sum_{f \in F} c_f = 0$.

DEFINITION 2: $\langle \mathbf{p}^*, c_1^*, \ldots, c_l^* \rangle$ is a *Marxian competitive equilibrium* iff there exists a decision $\{\tilde{\mathbf{x}}_1, \ldots, \tilde{\mathbf{x}}_l\}$ that satisfies (1)-(4) of Definition 1 with credits $c_1^*, \ldots, c_l^*$ and price system $\mathbf{p}^*$.

The aim of the following eleven, rather technical lemmas is to prepare the ground for the proof of the existence of a Marxian competitive equilibrium and, in this form, of a RGD.

LEMMA 1: *For every* $(\mathbf{p}, \xi_f) \in S \times \Xi_f$, $B_f(\mathbf{p}, \xi_f)$ *is nonempty, convex and compact. Analogously, for every* $(\mathbf{p}, \xi) \in S \times \Xi$, $B(\mathbf{p}, \xi)$ *is nonempty, convex and compact.*

Proof: Since $\xi_f \geq 0$ and $\tilde{\mathbf{0}} \in Y_f$, we have

$$\mathbf{p}\underline{\mathbf{0}} + \mathbf{pB0} = 0 \leq \xi_f$$

and so $\tilde{0} \in B_f(\mathbf{p}, \xi_f)$, which establishes that B_f is nonempty.

Let $\tilde{\mathbf{x}}, \tilde{\mathbf{y}} \in B_f(\mathbf{p}, \xi_f)$ and $\alpha \in [0, 1]$. Then $\mathbf{p}(\alpha \underline{\mathbf{x}}) + \mathbf{p}\mathbf{B}(\alpha \mathbf{x}) = \alpha(\mathbf{p}\underline{\mathbf{x}} + \mathbf{p}\mathbf{B}\mathbf{x}) \leq \alpha\xi_f$, $\mathbf{p}[(1 - \alpha)\underline{\mathbf{y}}] + \mathbf{p}\mathbf{B}[(1 - \alpha)\mathbf{y}] \leq (1 - \alpha)\xi_f$, and so

$$\mathbf{p}[\alpha\underline{\mathbf{x}} + (1 - \alpha)\underline{\mathbf{y}}] + \mathbf{p}\mathbf{B}[\alpha\mathbf{x} + (1 - \alpha)\mathbf{y}] \leq \xi_f$$

which proves that $\alpha\tilde{\mathbf{x}} + (1 - \alpha)\tilde{\mathbf{y}} \in B_f(\mathbf{p}, \xi_f)$.

Since the set $X = \{\mathbf{x} \mid \tilde{\mathbf{x}} \in B_f(\mathbf{p}, \xi_f)\}$ is bounded, because $\mathbf{p}\mathbf{B}\mathbf{x} \geq 0$ and $\mathbf{p}\mathbf{B}\mathbf{x} \leq \xi_f$ for every $\mathbf{x} \in X$, Lemma 1 of chapter 6 implies that $B_f(\mathbf{p}, \xi_f)$ is also bounded. In addition, $B_f(\mathbf{p}, \xi_f)$ is closed, because if $(\tilde{\mathbf{x}}_k)$ is a sequence in the same set converging to $\tilde{\mathbf{x}}$, then $(\mathbf{p}\underline{\mathbf{x}}_k + \mathbf{p}\mathbf{B}\mathbf{x}_k) \rightarrow \mathbf{p}\underline{\mathbf{x}} + \mathbf{p}\mathbf{B}\mathbf{x} \leq \xi_f$ with $\tilde{\mathbf{x}} \in Y_f$, which shows that $\tilde{\mathbf{x}} \in B_f(\mathbf{p}, \xi_f)$. Hence, $B_f(\mathbf{p}, \xi_f)$ is compact. The proof for $B(\mathbf{p}, \xi)$ is completely similar. $\square$

LEMMA 2: *For every* $(\mathbf{p}, \xi_f) \in S \times \Xi_f$, *the set* $A_f(\mathbf{p}, \xi_f)$ *is nonempty, convex, and compact. Analogously, the set* $A(\mathbf{p}, \xi)$ *is nonempty, convex, and compact for every* $(\mathbf{p}, \xi) \in S \times \Xi$.

Proof: Let $f : B_f(\mathbf{p}, \xi_f) \rightarrow \mathbf{R}$ be the function such that

$$f(\tilde{\mathbf{x}}) = [-\mathbf{p}\mathbf{B}, -\mathbf{p}, \mathbf{p}]\tilde{\mathbf{x}}^T = \mathbf{p}\hat{\mathbf{x}} - \mathbf{p}\mathbf{B}\mathbf{x}.$$

Since the inner product with a fixed vector is a continuous function, and $B_f(\mathbf{p}, \xi_f)$ is compact by Lemma 1, it follows that the set

$$\{\mathbf{p}\hat{\mathbf{x}} - \mathbf{p}\mathbf{B}\mathbf{x} \mid \tilde{\mathbf{x}} \in B_f(\mathbf{p}, \xi_f)\}$$

has a maximum, by Weierstrass' Theorem, and so $A_f(\mathbf{p}, \xi_f)$ is nonempty.

Let $\tilde{\mathbf{x}}, \tilde{\mathbf{y}} \in A_f(\mathbf{p}, \xi_f)$ and $\alpha \in [0, 1]$. Then $\mathbf{p}\hat{\mathbf{x}} - \mathbf{p}\mathbf{B}\mathbf{x} = \Pi_f(\mathbf{p}, \xi_f) = \mathbf{p}\hat{\mathbf{y}} - \mathbf{p}\mathbf{B}\mathbf{y}$, and so $\mathbf{p}[\alpha\hat{\mathbf{x}} + (1 - \alpha)\hat{\mathbf{y}}] - \mathbf{p}\mathbf{B}[\alpha\mathbf{x} + (1 - \alpha)\mathbf{y}] = \alpha[\mathbf{p}\hat{\mathbf{x}} - \mathbf{p}\mathbf{B}\mathbf{x}] + (1 - \alpha)[\mathbf{p}\hat{\mathbf{y}} - \mathbf{p}\mathbf{B}\mathbf{y}] = \alpha\Pi_f + (1 - \alpha)\Pi_f = \Pi_f$. This shows that $A_f(\mathbf{p}, \xi_f)$ is convex.

$A_f(\mathbf{p}, \xi_f)$ is bounded because it is contained in $B_f(\mathbf{p}, \xi_f)$, which is a compact set according to Lemma 2. It is obvious that any convergent sequence in $A_f(\mathbf{p}, \xi_f)$ converges to a maximizer in the same set. The proof for $A(\mathbf{p}, \xi)$ is completely similar. $\square$

A *correspondence* $\varphi : A \to B$ is a function $f : A \to \wp(B)$ from A to the power set of B that assigns to each element $x \in A$ a subset $\varphi(x) = f(x) \subseteq B$. A correspondence φ is said to be *lower semicontinuous* (lsc) if, whenever (x_k) is a sequence in A converging to x and $y \in \varphi(x)$, there is a sequence (y_k) converging to y such that $y_k \in \varphi(x_k)$ for every $k \in \omega$. Also, a correspondence φ is said to be *upper semicontinuous* (usc) if, whenever (x_k) is a sequence in A converging to a point x and (y_k) a sequence in B converging to y such that $y_k \in \varphi(x_k)$ for every $k \in \omega$, $y \in \varphi(x)$.

LEMMA 3: *For each $\xi_f \in \Xi_f$, $B_f : S \to Y_f$ is lsc. Analogously, $B : S \to Y$ is lsc for every $\xi \in \Xi$.*

Proof: Let $(\mathbf{p}_k)$ be a sequence in S converging to $\mathbf{p} \in S$, and let $\tilde{\mathbf{x}} \in B_f(\mathbf{p}, \xi_f)$. We have to show that there is a sequence $(\tilde{\mathbf{x}}_k)$ converging to $\tilde{\mathbf{x}}$ such that $\tilde{\mathbf{x}}_k \in B_f(\mathbf{p}_k, \xi_f)$ for every k. If $\tilde{\mathbf{x}} = \tilde{\mathbf{0}}$, we let $\tilde{\mathbf{x}}_k$ be the null vector for every k and the result trivially follows.

If $\tilde{\mathbf{x}}$ is nonnull and $\mathbf{p}\underline{\mathbf{x}} + \mathbf{p}\mathbf{B}\mathbf{x} > 0$ then, since $(\mathbf{p}_k\underline{\mathbf{x}} + \mathbf{p}_k\mathbf{B}\mathbf{x}) \to \mathbf{p}\underline{\mathbf{x}} + \mathbf{p}\mathbf{B}\mathbf{x}$, there is a positive integer N such that

$$k > N \quad \text{implies} \quad \mathbf{p}_k\underline{\mathbf{x}} + \mathbf{p}_k\mathbf{B}\mathbf{x} > 0.$$

Hence, defining

$$\lambda_k = \begin{cases} 0 & \text{if } \mathbf{p}_k\underline{\mathbf{x}} + \mathbf{p}_k\mathbf{B}\mathbf{x} = 0 \\ (\mathbf{p}\underline{\mathbf{x}} + \mathbf{p}\mathbf{B}\mathbf{x})/(\mathbf{p}_k\underline{\mathbf{x}} + \mathbf{p}_k\mathbf{B}\mathbf{x}) & \text{otherwise} \end{cases}$$

and $\tilde{\mathbf{x}}_k = \lambda_k\tilde{\mathbf{x}}$ we get

$$\mathbf{p}_k\underline{\mathbf{x}}_k + \mathbf{p}_k\mathbf{B}\mathbf{x}_k = \lambda_k[\mathbf{p}_k\underline{\mathbf{x}} + \mathbf{p}_k\mathbf{B}\mathbf{x}]$$
$$= \mathbf{p}\underline{\mathbf{x}} + \mathbf{p}\mathbf{B}\mathbf{x}$$
$$\leq \xi_f.$$

Thus, $\tilde{\mathbf{x}}_k \in B_f(\mathbf{p}_k, \xi_f)$ for every k. Moreover, it is clear that $(\lambda_k) \to 1$ and so $(\mathbf{x}_k) \to \mathbf{x}$.

Finally, if both $\mathbf{p}\underline{\mathbf{x}} = 0$ and $\mathbf{p}\mathbf{B}\mathbf{x} = 0$, let $(\tilde{\mathbf{y}}_k)$ be any sequence converging to $\tilde{\mathbf{x}}$. Then we have $(\mathbf{p}_k\underline{\mathbf{y}}_k) \to 0$ and $(\mathbf{p}_k\mathbf{B}\mathbf{y}_k) \to 0$.

Hence, there exists a positive integer N such that

$$k > N \Rightarrow \mathbf{p}_k \underline{\mathbf{y}}_k + \mathbf{p}_k \mathbf{B} \mathbf{y}_k < \xi_f$$
$$\Rightarrow \tilde{\mathbf{y}}_k \in B_f(\mathbf{p}_k, \xi_f).$$

The conclusion follows if we set

$$\tilde{\mathbf{x}}_k = \begin{cases} 0 & \text{if } \tilde{\mathbf{y}}_k \notin B_f(\mathbf{p}_k, \xi_f) \\ \tilde{\mathbf{y}}_k & \text{otherwise.} \end{cases}$$

An entirely similar argument establishes the proposition for the global function B. $\square$

LEMMA 4: (Berge's Maximum Theorem). *For each* $\xi_f \in \Xi_f$, $A_f : S \to Y_f$ *is usc. Analogously,* $A : S \to Y$ *is usc for every* $\xi \in \Xi$.

Proof: Let $(\mathbf{p}_k)$ be a sequence in S converging to $\mathbf{p}$, and let $(\tilde{\mathbf{x}}_k)$ be a sequence converging to $\tilde{\mathbf{x}}$ such that $\tilde{\mathbf{x}}_k \in A_f(\mathbf{p}_k, \xi_f)$ for every k. We have to show that $\tilde{\mathbf{x}} \in A_f(\mathbf{p}, \xi_f)$.

Since $\mathbf{p}_k \underline{\mathbf{x}}_k + \mathbf{p} \mathbf{B} \mathbf{x}_k \leq \xi_f$, it follows that

$$(\mathbf{p}_k \underline{\mathbf{x}}_k + \mathbf{p} \mathbf{B} \mathbf{x}_k) \to \mathbf{p} \underline{\mathbf{x}} + \mathbf{p} \mathbf{B} \mathbf{x} \leq \xi_f,$$

and so $\tilde{\mathbf{x}} \in B_f(\mathbf{p}, \xi_f)$.

In order to derive a contradiction, assume that $\tilde{\mathbf{x}} \notin A_f(\mathbf{p}, \xi_f)$. Then there must be a $\tilde{\mathbf{y}} \in B_f(\mathbf{p}, \xi_f)$ such that

$$\mathbf{p}\hat{\mathbf{y}} - \mathbf{p} \mathbf{B} \mathbf{y} = \gamma > \delta = \mathbf{p}\hat{\mathbf{x}} - \mathbf{p} \mathbf{B} \mathbf{x}.$$

It follows that the sequence $(\mathbf{p}_k \hat{\mathbf{x}}_k - \mathbf{p} \mathbf{B} \mathbf{x}_k)$ converges to δ and, since B_f is lsc, there is a sequence $(\tilde{\mathbf{y}}_k) \to \tilde{\mathbf{y}}$ such that $\tilde{\mathbf{y}}_k \in B_f(\mathbf{p}_k, \xi_f)$. Clearly, the sequence $(\mathbf{p}_k \hat{\mathbf{y}}_k - \mathbf{p} \mathbf{B} \mathbf{y}_k)$ converges to γ and so, for sufficiently large k, the profits that the operation of $\tilde{\mathbf{y}}_k$ would yield to firm f at prices $\mathbf{p}_k$ are larger than the profits that the operation of $\tilde{\mathbf{x}}_k$ by $f \in F$ would yield under the same prices. This contradicts the hypothesis that $\tilde{\mathbf{x}}_k \in A_f(\mathbf{p}_k, \xi_f)$, and so $\tilde{\mathbf{x}}$ must be in $A_f(\mathbf{p}, \xi_f)$. $\square$

LEMMA 5: *For every* $\tilde{\mathbf{x}} \in A(\mathbf{p}, \mathbf{ph})$, *there exists a decomposition* $\tilde{\mathbf{x}}_1, \ldots, \tilde{\mathbf{x}}_l$ *of* $\tilde{\mathbf{x}}$ *and credits* $c_1, \ldots, c_l$ *such that* $\tilde{\mathbf{x}}_f \in A_f(\mathbf{p}, c_f)$ *and* $\sum_{f \in F} c_f = 0$.

Proof: Let $\tilde{\mathbf{x}}_1, \ldots, \tilde{\mathbf{x}}_l$ be a decomposition of $\tilde{\mathbf{x}}$. Since

$$\mathbf{p}\underline{\mathbf{x}}_f + \mathbf{pBx}_f \leq \mathbf{ph},$$

it follows that there is a $\mathbf{c}_f \in C_f$ such that

$$\underline{\mathbf{x}}_f + \mathbf{Bx}_f = \mathbf{h}_f + \mathbf{c}_f.$$

Hence, $\xi_f = \mathbf{ph}_f + \mathbf{pc}_f = \mathbf{p}\underline{\mathbf{x}}_f + \mathbf{pBx}_f \in \Xi_f(\mathbf{p})$, and $\tilde{\mathbf{x}}_f \in \mathbf{B}_f(\mathbf{p}, \xi_f)$. If $\tilde{\mathbf{y}}_1, \ldots, \tilde{\mathbf{y}}_l$ are any processes with $\tilde{\mathbf{y}}_f \in A_f(\mathbf{p}, \xi_f)$, it follows that, for all $f \in F$,

$$\mathbf{p}\hat{\mathbf{y}}_f - \mathbf{pBy}_f \geq \mathbf{p}\hat{\mathbf{x}}_f - \mathbf{pBx}_f.$$

If $\tilde{\mathbf{x}}_f \notin A_f(\mathbf{p}, \xi_f)$ for some $f \in F$, we would have

$$\mathbf{p}\hat{\mathbf{y}} - \mathbf{pBy} = \mathbf{p}(\sum_{f \in F} \hat{\mathbf{y}}_f) - \mathbf{pB}(\sum_{f \in F} \mathbf{y}_f)$$
$$> \mathbf{p}\hat{\mathbf{x}} - \mathbf{pBx}.$$

Since $\mathbf{p}\underline{\mathbf{x}} + \mathbf{pBx} = \mathbf{ph}$, by Lemma 3 of chapter 6,

$$\mathbf{p}(\sum_{f \in F} \underline{\mathbf{y}}_f) + \mathbf{p}(\sum_{f \in F} \mathbf{y}) = \sum_{f \in F} \xi_f$$
$$= \sum_{f \in F} \mathbf{ph}_f + \sum_{f \in F} \mathbf{pc}_f$$
$$= \sum_{f \in F} (\mathbf{p}\underline{\mathbf{x}}_f + \mathbf{pBx}_f)$$
$$= \mathbf{ph},$$

it follows that $\tilde{\mathbf{y}} \in A(\mathbf{p}, \mathbf{ph})$, contradicting the hypothesis that $\tilde{\mathbf{x}} \in A(\mathbf{p}, \mathbf{ph})$. Finally, let $c_f = \mathbf{pc}_f$. Then $\sum_{f \in F} \mathbf{ph}_f + \sum_{f \in F} c_f = \sum_{f \in F} \mathbf{ph}_f$ implies $\sum_{f \in F} c_f = 0$. $\square$

Consider a set X and a correspondence $\varphi: X \to X$. A *fixed point* of the correspondence φ is an element $x^* \in X$ such that $x^* \in$

$\varphi(x^*)$. The main result in mathematical economics regarding the existence of fixed points is due to Kakutani. This shall be our next lemma here.

LEMMA 6: (Kakutani) *If X is a nonempty, compact, convex subset of* $\mathbf{R}^k$, *and if* φ *is an usc correspondence from X to X such that for all* $x \in X$ *the set* $\varphi(x)$ *is convex (nonempty), then* φ *has a fixed point.*

For a proof of this lemma, which is rather involved, the reader is referred to Kakutani (1941) or Nikaido (1968) in the Bibliography.

LEMMA 7: (Debreu). *Let* $\zeta: S \to Z$ *be a correspondence from S to the compact set* $Z \subseteq \mathbf{R}^m$. *If* ζ *is usc and, for all* $\mathbf{p} \in S$, $\zeta(\mathbf{p})$ *is nonempty, closed, convex, and* $\mathbf{p}\zeta(\mathbf{p}) \leq 0$, *then there is a* $\mathbf{p}$ *in S, and a* $\mathbf{z}$ *in* $\zeta(\mathbf{p})$, *such that* $\mathbf{z} \leq \mathbf{0}$.

Proof: In order to make use of Kakutani's theorem, a correspondence φ from $S \times Z$ into itself must be defined by means of the condition:

$$\varphi(\mathbf{p}, \mathbf{z}) = \mu(\mathbf{z}) \times \zeta(\mathbf{p}),$$

where $\mu: Z \to S$ is the correspondence that assigns to each $\mathbf{z} \in Z$ the set $\{\mathbf{p} \in S \mid \mathbf{p}$ maximizes $\mathbf{pz}$ on $S\}$. It is required to show that (i) $S \times Z$ is nonempty, convex and compact; (ii) φ is usc; and (iii) $\varphi(\mathbf{p}, \mathbf{z})$ is nonempty convex. When these conditions are fulfilled there exists a fixed point $(\mathbf{p}^*, \mathbf{z}^*) \in \mu(\mathbf{z}^*) \times \zeta(\mathbf{p}^*)$ and we have the following outcome.

$\mathbf{p}^* \in \mu(\mathbf{z}^*)$ implies that $\mathbf{pz}^* \leq \mathbf{p}^*\mathbf{z}^*$ for every $\mathbf{p} \in S$. $\mathbf{z}^* \in \zeta(\mathbf{p}^*)$ implies that $\mathbf{p}^*\mathbf{z}^* \leq 0$ and so $\mathbf{pz}^* \leq 0$ for every $\mathbf{p} \in S$. Since for every $k \in \{1, \ldots, m\}$ we can find a price system $\mathbf{p} \in S$ whose kth component, p_k is 1, while $p_j = 0$ for every other component $j \neq k$ ($1 \leq j \leq k$), it follows that $z_k^* \leq 0$ for every k, which is tantamount to $\mathbf{z}^* \leq \mathbf{0}$.

Thus, to show (i), since the Cartesian product of nonempty compact convex sets has also these attributes, it suffices to notice that Z is nonempty because S is nonempty and the correspondence ζ is nonempty valued. Z is by hypothesis compact and without loss of generality Z can be assumed to be convex, for otherwise it can be replaced by the closure of its convex hull without altering the final result.

φ is usc because the Cartesian product of usc correspondences is also usc and μ can be seen to be usc. In effect, let $(\mathbf{z}_k)$ be a sequence in Z converging to $\mathbf{z}$ for which there is a sequence $(\mathbf{p}_k)$ converging to $\mathbf{p}$ such that $\mathbf{p}_k \in \mu(\mathbf{z}_k)$ for every k. It is required to prove that $\mathbf{p} \in \mu(\mathbf{z})$. Let $\mathbf{p}^* \in \mu(\mathbf{z})$ and notice that, therefore, $\mathbf{p}^*\mathbf{z} \geq \mathbf{p}\mathbf{z} = \lim_{k\to\infty} \mathbf{p}_k\mathbf{z}_k$. Hence, since $\mathbf{p}_k\mathbf{z}_k \geq \mathbf{p}^*\mathbf{z}_k$,

$$\lim_{k\to\infty} \mathbf{p}_k\mathbf{z}_k \geq \lim_{k\to\infty} \mathbf{p}^*\mathbf{z}_k = \mathbf{p}^*\mathbf{z}.$$

It follows that $\mathbf{p}^*\mathbf{z} \geq \mathbf{p}\mathbf{z} \geq \mathbf{p}^*\mathbf{z}$ and so $\mathbf{p}\mathbf{z} = \mathbf{p}^*\mathbf{z}$, which means that $\mathbf{p}$ maximizes $\mathbf{p}\mathbf{z}$ on S, i.e. $\mathbf{p} \in \mu(\mathbf{z})$.

$\varphi(\mathbf{p}, \mathbf{z})$ is nonempty convex. This follows from the fact that $\zeta(\mathbf{p})$ is nonempty convex and that $\mu(\mathbf{z})$ can be shown to be so. Obviously, if $\mathbf{p}, \mathbf{p}' \in \mu(\mathbf{z})$ then $\mathbf{p}\mathbf{z} = \mathbf{p}'\mathbf{z}$ and so, for every $\alpha \in [0, 1]$,

$$(\alpha\mathbf{p} + (1 - \alpha)\mathbf{p}')\mathbf{z} = \alpha\mathbf{p}\mathbf{z} + (1 - \alpha)\mathbf{p}'\mathbf{z}$$
$$= \mathbf{p}\mathbf{z}$$

On the other hand,

$$\sum_{i=1}^{m}(\alpha p_i + (1 - \alpha)p_i') = \sum_{i=1}^{m}\alpha p_i + \sum_{i=1}^{m}(1 - \alpha)p_i'$$
$$= \alpha\sum_{i=1}^{m}p_i + (1 - \alpha)\sum_{i=1}^{m}p_i'$$
$$= \alpha + (1 - \alpha)$$
$$= 1.$$

This shows that $\alpha\mathbf{p} + (1 - \alpha)\mathbf{p}'$ is a price system in S that maximizes $\mathbf{p}\mathbf{z}$. Hence, it belongs to $\mu(\mathbf{z})$. This concludes the proof of Debreu's Theorem. $\square$

For every $\mathbf{p} \in S$, define the set

$$\zeta(\mathbf{p}) = \{\mathbf{Bx} - \hat{\mathbf{x}} \mid \tilde{\mathbf{x}} \in A(\mathbf{p}, \mathbf{ph})\}.$$

A correspondence is thus defined from the simplex S into the set $Z = \bigcup_{p \in S} \zeta(\mathbf{p})$. We have the following lemmas concerning ζ and Z.

LEMMA 8: $\zeta(\mathbf{p})$ *is nonempty for every* $\mathbf{p} \in S$.

Proof: By Lemma 3, $A(\mathbf{p}, \xi)$ is nonempty for every $\xi \in \Xi$, in particular for $\mathbf{ph} \in \Xi$. Hence, for all $\tilde{\mathbf{x}} \in A(\mathbf{p}, \mathbf{ph})$, $\mathbf{Bx} - \hat{\mathbf{x}} \in \zeta(\mathbf{p})$. $\square$

LEMMA 9: Z *is compact.*

Proof: Let A be the set $\bigcup_{p \in S} A(\mathbf{p}, \mathbf{ph})$. Then Z is the image of A under the continuous function $f : A \rightarrow Z$, where $f(\tilde{\mathbf{x}}) = \mathbf{Bx} - \hat{\mathbf{x}}$. Since the image of a compact set in a linear space is also compact, it will suffice to show that A is compact.

Since S is bounded, $\{\mathbf{ph} \mid \mathbf{p} \in S\}$ is also bounded and so it has a supremum μ. It follows that if $\tilde{\mathbf{x}} \in A$, then $\mathbf{pBx} \leq \mathbf{ph} \leq \mu$ and so $\{\mathbf{x} \mid \tilde{\mathbf{x}} \in A\}$ is bounded.

If $(\tilde{\mathbf{x}}_k)$ is any sequence in A converging to $\tilde{\mathbf{x}}$, for each k select a price $\mathbf{p}_k \in S$ such that $\mathbf{x}_k \in A(\mathbf{p}_k, \mathbf{p}_k \mathbf{h})$. Since S is compact, without loss of generality we may suppose that $(\mathbf{p}_k)$ converges to a price $\mathbf{p} \in S$. But, by Lemma 5, $A : S \rightarrow Y$ is usc and therefore $\tilde{\mathbf{x}} \in A(\mathbf{p}, \mathbf{ph}) \subseteq A$, which establishes that A is also closed. $\square$

LEMMA 10: ζ *is usc.*

Proof: Let $\mathbf{p}$ be a point of S and $(\mathbf{p}_k)$ a sequence in S converging to $\mathbf{p}$. Let $(\mathbf{z}_k)$ be a sequence in Z converging to $\mathbf{z}$ such that $\mathbf{z}_k \in \zeta(\mathbf{p}_k)$. I will show that $\mathbf{z} \in \zeta(\mathbf{p})$.

Each $\mathbf{z}_k$ is of the form

$$\mathbf{z}_k = \mathbf{Bx}_k - \hat{\mathbf{x}}_k$$

where $\tilde{\mathbf{x}}_k \in A(\mathbf{p}_k, \mathbf{p}_k \mathbf{h})$. It follows that

$$\mathbf{p}_k \mathbf{Bx}_k \leq \mu$$

where μ is as given in the proof of Lemma 9, and so the set of vectors of labor inputs $\{\mathbf{x}_k\}$ of the terms in the sequence $(\mathbf{x}_k)$ is bounded. Thus, by Lemma 1 of chapter 6 this sequence itself is bounded and so we can assume that converges to a process $\tilde{\mathbf{x}}$. Clearly, the limit $\mathbf{z}$ of $(\mathbf{z}_k)$ must be equal to $\mathbf{Bx} - \hat{\mathbf{x}}$. Since the correspondence A is usc, it follows that $\tilde{\mathbf{x}} \in A(\mathbf{p}, \mathbf{ph})$. This establishes that $\mathbf{z} \in \zeta(\mathbf{p})$. $\square$

LEMMA 11: *For every* $\mathbf{p} \in S$, $\zeta(\mathbf{p})$ *is closed, convex, and* $\mathbf{p}\zeta(\mathbf{p}) \leq 0$.

Proof: Let $(\mathbf{z}_k)$ be a sequence in $\zeta(\mathbf{p})$ converging to $\mathbf{z}$. In order to show that $\mathbf{z} \in \zeta(\mathbf{p})$, notice that $\mathbf{z}_k$ is of the form

$$\mathbf{z}_k = \mathbf{Bx}_k - \hat{\mathbf{x}}_k$$

for processes $\tilde{\mathbf{x}}_k \in A(\mathbf{p}, \mathbf{ph})$. Thus, $\mathbf{pBx}_k - \mathbf{p}\hat{\mathbf{x}}_k \leq \mathbf{ph}$, which implies that $(\tilde{\mathbf{x}}_k)$ is bounded. Hence, we may infer that $(\tilde{\mathbf{x}}_k)$ converges to $\tilde{\mathbf{x}}$, and likewise that $(\mathbf{z}_k) \rightarrow \mathbf{Bx} - \hat{\mathbf{x}}$. Since $A(\mathbf{p}, \mathbf{ph})$ is closed, $\tilde{\mathbf{x}} \in A(\mathbf{p}, \mathbf{ph})$, and so $\mathbf{z} \in \zeta(\mathbf{p})$.

In order to show that $\zeta(\mathbf{p})$ is convex, let $\mathbf{z} = \mathbf{Bx} - \hat{\mathbf{x}}$ and $\mathbf{u} = \mathbf{By} - \hat{\mathbf{y}}$ be any elements of $\zeta(\mathbf{p})$ and $\alpha \in [0, 1]$. Then

$$\alpha\mathbf{z} + (1 - \alpha)\mathbf{u} = \mathbf{B}(\alpha\mathbf{x} + (1 - \alpha\mathbf{y})) - (\alpha\hat{\mathbf{x}} + (1 - \alpha)\hat{\mathbf{y}})$$

where $\tilde{\mathbf{x}}, \tilde{\mathbf{y}} \in A(\mathbf{p}, \mathbf{ph})$. Since this set is convex, $\alpha\tilde{\mathbf{x}} + (1 - \alpha)\tilde{\mathbf{y}} \in A(\mathbf{p}, \mathbf{ph})$ and so $\alpha\mathbf{z} + (1 - \alpha)\mathbf{u} \in \zeta(\mathbf{p})$.

Finally, since $\mathbf{p}\underline{0} + \mathbf{B0} \leq \mathbf{ph}$, no element $\tilde{\mathbf{x}} \in A(\mathbf{p}, \mathbf{ph})$ is such that $\mathbf{p}\hat{\mathbf{x}} - \mathbf{pBx} < 0$. Hence, for every $\mathbf{z} = \mathbf{Bx} - \hat{\mathbf{x}} \in \zeta(\mathbf{p})$, $-\mathbf{pz} = \mathbf{p}\hat{\mathbf{x}} - \mathbf{pBx} \geq 0$. This proves that $-\mathbf{p}\zeta(\mathbf{p}) \geq 0$ and, therefore, $\mathbf{p}\zeta(\mathbf{p}) \leq 0$. $\square$

THEOREM: *There exists a Marxian competitive equilibrium.*

Proof: From lemmas 7 and 11 follows that there is a $\mathbf{p}$ in S and a $\mathbf{z}$ in $\zeta(\mathbf{p})$ such that $\mathbf{z} \leq \mathbf{0}$. In other words, there is a process $\tilde{\mathbf{x}} \in A(\mathbf{p}, \mathbf{ph})$ such that $\mathbf{Bx} - \hat{\mathbf{x}} \leq \mathbf{0}$, i.e. $\hat{\mathbf{x}} \geq \mathbf{Bx}$. This establishes that $\tilde{\mathbf{x}}$ is reproducible. Moreover, since $\mathbf{h} > \mathbf{0}$ and $\mathbf{p} \geq \mathbf{0}$, $\mathbf{ph} > 0$. On the other hand, $\tilde{\mathbf{x}} \in A(\mathbf{p}, \mathbf{ph})$ implies that $\mathbf{p}\underline{\mathbf{x}} + \mathbf{pBx} = \mathbf{ph} > 0$ and so $\tilde{\mathbf{x}}$ is nonnull.

By Lemma 5, there exists a decomposition $\tilde{\mathbf{x}}_1, \ldots, \tilde{\mathbf{x}}_l$ of $\tilde{\mathbf{x}}$ and credits $c_1, \ldots, c_l$ such that $\tilde{\mathbf{x}}_j \in A_j(\mathbf{p}, c_j)$ and $\sum_{j \in F} c_j = 0$.

Finally, since $\mathbf{p}\underline{\mathbf{x}} + \mathbf{p}\mathbf{B}\mathbf{x} \leq \mathbf{p}\mathbf{h}$, Axiom (5) of Definition 1 in chapter 6 implies that $\underline{\mathbf{x}} + \mathbf{B}\mathbf{x} \leqq \mathbf{t}$. $\square$

COROLLARY: *There exists a global reproducible solution.*

Proof: Immediate from the Theorem.

In this form we conclude the proof of the existence of a RGD. It follows that it is possible to make compatible the global reproducibility of the economy with the pursuing of self-interest by the firms. On the other hand, it is easy to see that the firms might choose production processes that do not necessarily reproduce the economy. Supposedly, the law of Supply and Demand should eventually lead the firms to choose precisely those processes that reproduce the economy, but this fact (if it is indeed a fact) is not reflected in the arguments provided here.

Chapter 8

THE PROTOTYPE REVISITED

After the long disquisition that we had to carry on in order to cope with the difficulties raised against the prototype of MTV, we are at last in an advantageous position to solve such difficulties. In the present chapter I will present, in a very detailed way, a reconstruction of the Leontief model of MTV with heterogeneous labor. This model shall appear here as a model of MTV in the logical sense, i.e. as a Marxian capitalist economy, albeit one with very special (and mathematically nice) properties.

In the first part of the chapter I will discuss again the "intended interpretation" of the mathematical theory. This discourse is more than a mere formality: it intends to provide the mathematical formulas with an economic meaning. It is just a fact that the Leontief economy described thereby is a Meinongian economy that has no existence whatsover. At any rate, *that* is the economy to which the concepts and equations of the theory apply, and real market economies are only roughly approximated by that model. I shall not discuss here what Professor Samuelson calls the F-twist, namely, that

> It is a positive merit of a theory that (some of) its content and assumptions be unrealistic, since only if it is not tailored closely to one small bit of reality can it give a useful fit to a wide spread of empirical situations. Unless we explain complex reality by something simpler than itself we have accomplished little (period or by theorizing).[1]

Yet, it must be said that the problem of the Marx-Leontief model is precisely that it fits too closely a small piece of irreality (!) and not at all the capitalist economies in their full universality. It is an interesting problem in the philosophy of economics the usefulness of idealized models, a problem that I discussed to some extent in chapter 4.

In the second section of this chapter I will present the Leontief economy as a mathematical structure satisfying the set-theoretic predicate introduced in Definition 4 of chapter 6 ('Marxian capitalist economy'). After discussing the special properties of the RGD for this economy, I will prove that in this type of economies holds a very strong form of the Law of Value and that, even though the explotation rate might not be unique, yet there is an explicit function correlating these rates with the profit rate (which *is* unique). I conclude the chapter discussing the conditions under which full employment in the economy is guaranteed.

8.1 THE LEONTIEF ECONOMY

A Leontief capitalist economy is pretty much like the simple Marxian economy of chapter 1, except that here the firms have a whole set of production possibilities to choose and labor is not assumed to be homogeneous.

As in the simple economy of chapter 1, at the beginning of an economic cycle the firms are endowed with certain initial holdings of capital and wage goods. Together with their aim to make the greatest possible profit (this is what makes each one of them behave like an *homo œconomicus*), they desire to maintain the economic system indefinitely, to keep making money in the future. Hence, they also desire the *reproduction* (be it simple or widened) of the system (this is what makes each one of them behave like a responsible *homo politicus*). A very interesting question that arises then is whether the capitalist can be simultaneously an *homo œconomicus* and an *homo politicus*, whether the firms can have their cake and also eat it. We saw on chapter 7 that the answer is affirmative, but we shall see here how these two ends can be coordinated in a Leontief economy.

Unlike the simple model of chapter 1, in the present one the behavior of the firms is considered in an explicit way. We have l firms in a set F and the production possibilities set of any firm $f \in F$ is assumed to have very special properties (this is what makes the economy to be of Leontief type). The first one of these special properties is that the production processes are able to produce only one kind of good (there is no joint production), and each firm is assumed to be a specialist in the production of one type of commodity. Another one is that there are no alternative technologies, in the sense that all production processes use the same types of labor and means of production, even though different positive combinations of the same are allowed. Hence, "the same" technology can be operated at different levels of efficiency.

There are also properties pertaining to the whole. It is assumed that each of the m kinds of goods can be produced by some firm. Hence, if the price system is feasible, every good shall be produced within the economy. It can be proven that there exists at least one semiproductive technology, i.e. that the technology has sufficient development as to be able to sustain the reproduction of the economy. Quite another matter is whether the "enlightened self-interest" of the firms guarantees that the actual technology is reproducible. It is shown that there exist reproducible global decisions (RGDs), i.e. global decisions of the firms that maximize the profit of each one and which also are feasible and reproducible. If the firms choose one of these RGDs, not only the economy is reproducible, but also it is possible to prove that there is only one, strictly positive equilibrium price system for the economy. This in turn entails a very strong form of the Law of Value, which makes it possible to assign a unique positive value to each commodity in the system. Finally, it is shown that if the initial holdings of the firms are sufficiently large then full employment and reproduction is guaranteed.

8.2 MATHEMATICAL MODELING OF THE LEONTIEF ECONOMY

In a Marxian capitalist economy, consider a convex polyhedral cone (a technology) X contained in the aggregate set Y and let

$\tilde{\mathbf{x}}_1, \ldots, \tilde{\mathbf{x}}_m$ be a set of linearly independent processes spanning X. Assume that there is no joint production in these processes and also no alternative techniques. I shall call a set X with these properties a Leontief technology. More precisely:

DEFINITION 1: Let Y be the aggregate set of possibilities of production of a Marxian economy. A *Leontief technology* is a convex polyhedral cone X spanned by a finite set $\tilde{\mathbf{x}}_1, \ldots, \tilde{\mathbf{x}}_m$ of elements of Y having the properties that none of these processes yields joint products and there are no alternative techniques, i.e. $\bar{x}_{ij} > 0$ iff $i = j$ $(i, j = 1, \ldots, m)$.

Following Marx's distinction into sectors, it is possible to distinguish among the types of goods produced by X two main kinds, namely, capital goods and wage goods. There is no loss of generality in supposing that these two classes are disjoint, and so we are justified in introducing the following definition.

DEFINITION 2: A good of type i $(i = 1, \ldots, m)$ is called a *capital good* in X if there is a process $\tilde{\mathbf{x}} \in X$ such that $\underline{x}_i > 0$. A good of type j $(j = 1, \ldots, m)$ is called a *wage good* if there is a process $\tilde{\mathbf{x}} \in X$ such that $\bar{x}_j = b$, where b is a positive entry in the matrix $\mathbf{B}$ of consumption bundles for the working class. Any good which is not a wage or a capital good is called a *luxury good*.

In order to avoid unnecessary complications, I shall assume that there are no luxury goods (assuming that there are just adds more details to the proofs, as in chapter 1). Hence, by convention, the types of commodities can be rearranged in such a way that those of type $1, \ldots, k$ are capital goods and those of type $k + 1, \ldots, m$ are wage goods. With this convention, the matrix of wage and capital goods industries can be written as

$$
\mathbf{X} = \begin{bmatrix}
\underline{x}_{11} & \cdots & \underline{x}_{1k} & \underline{x}_{1k+1} & \cdots & \underline{x}_{1m} \\
\vdots & & \vdots & \vdots & & \vdots \\
\underline{x}_{k1} & \cdots & \underline{x}_{kk} & \underline{x}_{kk+1} & \cdots & \underline{x}_{km} \\
0_{k+11} & \cdots & 0_{k+1k} & 0_{k+1k+1} & \cdots & 0_{k+1m} \\
\vdots & & \vdots & \vdots & & \vdots \\
0_{m1} & \cdots & 0_{mk} & 0_{mk+1} & \cdots & 0_{mm}
\end{bmatrix}.
$$

The matrix of wage goods can be written as

$$\mathbf{B} = \begin{bmatrix} 0_1 & \cdots & 0_1 \\ \vdots & & \vdots \\ 0_k & \cdots & 0_k \\ b_{k+11} & \cdots & b_{k+1n} \\ b_{m1} & \cdots & b_{mn} \end{bmatrix}$$

If we let

$$\mathbf{L} = \begin{bmatrix} x_{11} & \cdots & x_{1m} \\ \vdots & & \vdots \\ x_{n1} & \cdots & x_{nm} \end{bmatrix}$$

be the matrix of labor inputs corresponding to the wage and capital good industries, then the wage and capital inputs matrix is the sum $\mathbf{D} = \mathbf{X} + \mathbf{BL}$. I shall call a matrix like this a *global matrix*.

I am interested in defining a concept of interconnection among the capital and wage good industries. This concept has been already introduced in chapter 1, where I discussed the concept of interconnectedness of the wage and capital goods industries of the prototype, represented by matrix $\mathbf{C}$. I want also to define this notion, in an analogous way, for $\mathbf{D}$, but I also want to discuss a little bit more its economic meaning and the adequacy of its mathematical representation.

Imagine an economy that can be severed into two groups of industries, such that the industries in any of these two groups do not need as wage or capital inputs goods of the type produced by the industries in the other group. This is an economy which is not interconnected if, indeed, it can be called *one* economy at all. Let $i_1, \ldots i_n$ be the industries in the first group, and $i_{n+1}, \ldots, i_m$ those in the second. The arbitrary entry d_{ij} of matrix $\mathbf{D}$ is the amount of output of industry i consumed by industry j either as a wage or a capital good. Clearly, the lack of dependency of industries $i_1, \ldots i_n$ with respect to industries $i_{n+1}, \ldots, i_m$ means that no good produced by these is consumed by the former, and so $d_{ij} = 0$ whenever $j = i_1, \ldots i_n$ and $i = i_{n+1}, \ldots, i_m$. Hence, one way of expressing the interrelatednes of the technology is to demand that the matrix $\mathbf{D}$ be indecomposable, which is tantamount to preclude the existence of such subsets of indices. Another important property of the technology $\mathbf{D}$ was mentioned in connection with the prototype in chapter 1; this is its semiproductivity, a purely technological property without which no economy is feasible. I shall introduce these two properties together in the following definition.

DEFINITION 3: Let X be a Leontief technology in a Marxian capitalist economy. We say that X is *interconnected* iff the corresponding global matrix $\mathbf{D}$ is indecomposable. Also, we say that X is *semiproductive* iff $\mathbf{Dy} \leq \mathbf{y}$ for some positive (column) vector $\mathbf{y}$. If some Leontief technology contained in the aggregated set Y is interconnected, the set Y itself is said to be also *interconnected*. If some Leontief technology contained in Y is semiproductive, the set Y is also said to be semiproductive.

The production possibilities sets of a Marxian economy are closed convex cones with certain properties. An additional property that these sets may, or may not have, is that of being technologies apt to produce only one kind of good. This possibility motivates the concept of a specialized firm, which is introduced at this point.

DEFINITION 4: Let F be the set of firms of a Marxian capitalist economy. A firm $f \in F$ is called a *specialist* iff its production possibility set produces a unique type of good. That is to say, there are in the economy precisely m production possibilities sets Y_f, firm f has access only to one of these, and for this Y_f there is a positive integer i ($1 \leq i \leq m$) such that $\bar{x}_i > 0$ for some $\tilde{\mathbf{x}} \in Y_f$ and $\bar{x}_i \geq 0$ for every $\tilde{\mathbf{x}} \in Y_f$, whereas $\bar{x}_j = 0$ if $j \neq i$.

On top of precluding joint production, the production possibilities sets may also shun alternative technologies, in the sense of allowing different proportions of inputs and outputs but avoiding the use of different types of inputs. In other words, these sets may allow different degrees of efficiency to produce one kind of commodity, but they do not allow techniques using essentially different types of labor and/or capital goods. This idea is more precisely defined as follows.

DEFINITION 5: The set Y_f is said to *preclude alternative technologies* if the following two conditions are fulfilled: (1) If there is a process $\tilde{\mathbf{x}} \in Y_f$ such that $x_i = 0$, then $y_i = 0$ for every $\tilde{\mathbf{y}} \in Y_f$. (2) If there is a process $\tilde{\mathbf{x}} \in Y_f$ such that $\underline{x}_i = 0$, then $\underline{y}_i = 0$ for every $\tilde{\mathbf{y}} \in Y_f$. If in a Marxian economy all production possibilities sets preclude alternative technologies, we say that *there are no alternative technologies* in that economy.

It can be proven that if there are no alternative technologies in an economy and there is at least one interconnected Leontief technology included in Y, i.e. if Y is interconnected, then every Leontief technology included in Y is interconnected. This is the first lemma of the chapter.

LEMMA 1: *If in a Marxian economy there are no alternative technologies, every good can be produced by some firm, and every firm is a specialist, then there is one interconnected Leontief technology included in Y iff every Leontief technology included in Y is interconnected.*

Proof: Suppose that there is one interconnected Leontief technology $X \subseteq Y$. By Definition 1, this set is spanned by a finite set of processes $\tilde{\mathbf{x}}_1, \ldots, \tilde{\mathbf{x}}_m$ of elements of Y such that $\bar{x}_{ij} > 0$ iff $i = j$ $(i, j = 1, \ldots, m)$. Since every firm is a specialist, each production possibility set Y_f produces a unique type of good. If there are more firms than types of goods $(l > m)$ some of these production possibilities sets Y_f must be shared by two or more firms. At any rate, there cannot be less production sets than firms, because every good can be produced by some firm. Hence, it follows that there are exactly m production possibilities sets $Y_1, \ldots Y_m$ in the economy. By definition of Y, process $\tilde{\mathbf{x}}_i$ must then belong to set Y_i, even though two or more firms could have chosen different processes belonging to this set: in such a case $\tilde{\mathbf{x}}_i$ is just the aggregation of these processes. By assumption there are no alternative technologies, and so all process are similar in having zeros (if at all) in the same entries always. Therefore, the aggregate process $\tilde{\mathbf{x}}_i$ is similar to any of the elements of the set where its components were chosen (in fact, it is a member of this set, which is a cone) and so the global matrix of X is similar to the global matrix of any other Leontief technology included in Y. It follows that if the global matrix of X is indecomposable then all such matrices are indecomposable. $\square$

The following lemma establishes the equivalence of the notation for production processes in terms of $2m + n$ vectors with the representation of the same in terms of global matrices. Usually, the Leontief technology is described in terms of these matrices but, as the lemma asserts, there is an alternative description in terms of a convex polyhedral cone.

LEMMA 2: *Every element* $\tilde{\mathbf{x}}$ *in the Leontief technology* $X \subseteq Y$ *can be represented as* $\tilde{\mathbf{x}} = [\mathbf{Ls}, \mathbf{Xs}, \mathbf{s}]$, *for some state* $\mathbf{s} \geq \mathbf{0}$.

Proof: Let $\tilde{\mathbf{x}}_1, ..., \tilde{\mathbf{x}}_m$ be linearly independent processes spanning X, chosen in such a way that $\bar{x}_{ii} = 1$. Let $\mathbf{L} = [\mathbf{x}_1^T \cdots \mathbf{x}_m^T]$ be the matrix whose columns are the vectors of labor inputs of these processes, let $\mathbf{X} = [\underline{\mathbf{x}}_1^T \cdots \underline{\mathbf{x}}_m^T]$, and let $\tilde{\mathbf{x}}$ be any element of X. Then there exist unique nonnegative real numbers $\alpha_1, ..., \alpha_m$ such that $\tilde{\mathbf{x}} = \alpha_1 \tilde{\mathbf{x}}_1 + \cdots + \alpha_m \tilde{\mathbf{x}}_m$. Let $\bar{\mathbf{x}}_i$ be the state that has 1 at place i and zeros everywhere else and notice that $\bar{\mathbf{x}} = \sum_{i=1}^{m} \alpha_i \bar{\mathbf{x}}_i$. Then $\tilde{\mathbf{x}}_i = [\mathbf{L}\bar{\mathbf{x}}_i, \mathbf{X}\bar{\mathbf{x}}_i, \bar{\mathbf{x}}_i]$ and so[2]

$$\tilde{\mathbf{x}} = [\sum_{i=1}^{m} \alpha_i \mathbf{L}\bar{\mathbf{x}}_i, \sum_{i=1}^{m} \alpha_i \mathbf{X}\bar{\mathbf{x}}_i, \sum_{i=1}^{m} \alpha_i \bar{\mathbf{x}}_i]$$

$$= [\mathbf{L}\bar{\mathbf{x}}, \mathbf{X}\bar{\mathbf{x}}, \bar{\mathbf{x}}]. \quad \square$$

In terms of the newly defined concepts, it is possible to introduce the fundamental definition of the chapter. This is the definition of a Leontief capitalist economy, which is a Marxian economy with special properties.

DEFINITION 6: A *Leontief capitalist economy* is a Marxian capitalist economy $\mathfrak{L}$ in which every firm is a specialist, every type of good can be produced by some firm, there are no alternative technologies, and Y is interconnected.

If the price system is feasible, as a result of the decision of the firms, l processes $\tilde{\mathbf{x}}_1, \ldots \tilde{\mathbf{x}}_l$ are chosen, all of these processes being efficient. The convex cone P spanned by these processes (the actual technology of $\mathfrak{L}$) is in fact interconnected, a result that is stated as the first theorem of this chapter.

THEOREM 1: *In a Leontief economy the actual technology is always interconnected, whenever the price system is feasible.*

Proof: If the price system is feasible, then every firm is encouraged to obtain positive profits by operating some nonnull process. Hence, by Lemma 1, the technology chosen by the firms is interconnected. $\square$

We know from chapter 6 that if the price system is feasible then every firm operates a nonnull process, and these processes are chosen in such a way that the resulting price turns out to be an equilibrium price, i.e. a price at which the profit rate is the same for all the production processes in the economy (recall that this is due to the existence of a financial capital market). Quite another question is whether the behavior of the firms, by itself, guarantees that the processes they decide to operate can be reproduced, in the sense of being able to produce the capital and wage goods they consume and perhaps obtain some excedents. A social and historical presupposition of the starting of a new production cycle in a market economy is the previous existence of enough resources, as well as sufficiently efficient technologies in all branches of the economy, as to guarantee, at the very least, the possibility of a "simple reproduction" of the economy; that is, the possibility of replacing the worn out capital goods, and of "reproducing" that part of the working class which is employed, that is to say, of providing for the demand **B** of those who are employed. The existence of a RGD means that there is a technology which, if operated at an appropiate level, is able to reproduce the capital goods and the employed labor power of the economy. Naturally, the condition required to guarantee that all the firms operate a reproducible nonnull process is the feasibility of the equilibrium price system.

THEOREM 2: *In a Leontief economy there is at least one RGD. If the equilibrium price corresponding to this RGD is feasible, then the global matrix corresponding to this RGD is semiproductive.*

Proof: Since a Leontief economy is a Marxian capitalist economy, the results of chapter 7 imply that there is a RGD $\{\tilde{x}_1, \ldots, \tilde{x}_m\}$. Let $\mathbf{D}$ be the global matrix corresponding to this technology. Since all firms operate a nonnull process, because the corresponding equilibrium price system is feasible, the columns of $\mathbf{D}$ are all nonnegative and, moreover, the output $\bar{x}$ of the global process is positive. It is easy to see that the condition of reproducibility in Definition 1 of chapter 7, that $\bar{x} \geq x + \mathbf{B}x$, is equivalent to $\mathbf{D}\bar{x} \leq \bar{x}$, i.e. to the semiproductivity of $\mathbf{D}$. $\square$

A very curious and special trait of Leontief economies is the fact that, for any of the actual technologies generated by a RGD with a feasible price, there is only one uniform profit rate and only one equilibrium price (up to multiplication by a positive scalar). This is due to the interconnectedness and semiproductivity of the industries, a property that —as we saw— in these economies can be formulated in terms of the indecomposability and semiproductivity of the global matrices. This result shall be the next theorem of the chapter.

THEOREM 3: *There is a unique (up to multiplication by a positive scalar) uniform profit rate for the actual technology of $\mathfrak{L}$, whenever the decision of the firms is a RGD with a feasible price system. Moreover, there is also a unique (up to multiplication by a positive scalar) system of prices at which the uniform profit rate obtains. This system of prices is positive.*

Proof: Let **D** be the global matrix associated to the actual technology chosen by the firms (a RGD). By theorems 1 and 2, we know that this technology is interconnected and semiproductive, and so **D** is indecomposable and semiproductive.

By the indecomposability of matrix **D** and the Perron-Frobenius theorem, there is a unique positive real eigenvalue γ (the Frobenius root) to which there corresponds a unique (up to multiplication by a positive scalar) positive left eigenvector **p**:

$$\mathbf{p}\mathbf{D} = \mathbf{p}\gamma.$$

Let

$$\pi = \frac{1 - \gamma}{\gamma}.$$

Since **D** is semiproductive, there is a positive vector **y** such that $\mathbf{D}\mathbf{y} \leqq \mathbf{y}$. Hence, $\mathbf{p}\mathbf{D}\mathbf{y} \leq \mathbf{p}\mathbf{y}$ and so $\gamma \leq 1$. It follows that π is a uniquely determined nonnegative number and that $1 + \pi = \gamma^{-1}$. Therefore,

$$\mathbf{p} = (1 + \pi)\mathbf{p}\mathbf{D}. \qquad \square$$

I shall call this price **p** the *equilibrium price* (EP), very much as in chapter 1, and the profit rate π shall be called the *equilibrium profit rate* (EPR). It can be shown that the profit rate as introduced in Definition 4 of chapter 5 coincides with the number π above. That is:

THEOREM 4: *For every $\tilde{\mathbf{x}} \in P$, the profit rate of $\tilde{\mathbf{x}}$ is identical to the EPR π.*

Proof: For any $\tilde{\mathbf{x}} \in P$,

$$\begin{aligned}
\mathbf{p}\bar{\mathbf{x}} &= (1 + \pi)\mathbf{p}\mathbf{D} \\
&= (1 + \pi)[\mathbf{p}\mathbf{X}\bar{\mathbf{x}} + \mathbf{p}(\mathbf{B}\mathbf{L}\bar{\mathbf{x}})] \\
&= (1 + \pi)[\mathbf{p}\underline{\mathbf{x}} + \mathbf{w}\mathbf{x}]
\end{aligned}$$

and so

$$0 \leq \pi = \frac{\mathbf{p}\hat{\mathbf{x}} - \mathbf{w}\mathbf{x}}{\mathbf{p}\underline{\mathbf{x}} + \mathbf{w}\mathbf{x}}.$$

Thus, π is in fact the profit rate of any labor process in P. $\square$

By virtue of Theorem 1 of chapter 6, we know that the profit rate is uniform and identical to the interest rate, and also that the Law of Value holds in a Leontief economy. It follows also that the EP π is identical to the interest rate prevailing in the financial capital market. The next theorem is a very strong version of the Law of Value, since it establishes that the equilibrium prices are in fact the unique labor-values of the commodities.

THEOREM 5: *In a Leontief economy, suppose that the technology chosen by the firms is a RGD, and let $l_i = \mathbf{r}\mathbf{x}_i$ be the amount of abstract live labor expended in the production of one unit of good i ($i = 1, \ldots, m$). Then the EP is in fact a system of labor values, in the sense that it satisfies the equation*

$$p_i = \underline{x}_{1i}p_1 + \cdots + \underline{x}_{ki}p_k + l_i$$

for every kind of good i ($= 1, \ldots, m$). As a matter of fact, prices are the only labor values in this economy, up to scale transformations.

Proof: By Definition 8 of chapter 5, $\lambda(\hat{\mathbf{x}}_i) = \mathbf{r}\mathbf{x}_i$, where $\mathbf{r}$ is a positive reduction of labor. Hence,

$$l_i = \mathbf{p}\hat{\mathbf{x}}_i$$
$$= \sum_{j=1}^{m} p_j \bar{x}_j i - \sum_{j=1}^{m} p_j \underline{x}_j i$$
$$= p_i \bar{x}_{ii} - \sum_{j=1}^{m} p_j \underline{x}_{ji}$$
$$= p_i - \sum_{j=1}^{m} p_j \underline{x}_{ji}.$$

Thus,

$$p_i = \underline{x}_{1i}p_1 + \cdots + \underline{x}_{ki}p_k + l_i.$$

Let $\mathbf{A}_I$ and $\mathbf{A}_{II}$ be, as in chapter 1, the matrices

$$\mathbf{A}_I = [\underline{\mathbf{x}}_1^T \cdots \underline{\mathbf{x}}_k^T]$$

and

$$\mathbf{A}_{\mathrm{II}} = [\underline{\mathbf{x}}_{k+1}^{\mathrm{T}} \cdots \underline{\mathbf{x}}_m^{\mathrm{T}}]$$

of capital goods and wage goods industries, respectively. Let $\mathbf{L}_\mathrm{I}$ and $\mathbf{L}_\mathrm{II}$ be the matrices given by conditions

$$\mathbf{L}_\mathrm{I} = [l_1 \cdots l_k] \quad \text{and} \quad \mathbf{L}_\mathrm{II} = [l_{k+1} \cdots l_m].$$

Since $\mathbf{A}_\mathrm{I}$ is quasiproductive and indecomposable, Theorem 1 of chapter 1 implies that the matrices

$$\Lambda_\mathrm{I} = \Lambda_\mathrm{I}\mathbf{A}_\mathrm{I} + \mathbf{L}_\mathrm{I} \qquad \text{and} \qquad \Lambda_\mathrm{II} = \Lambda_\mathrm{I}\mathbf{A}_\mathrm{II} + \mathbf{L}_\mathrm{II}$$

are unique and positive, where

$$\Lambda_\mathrm{I} = [\lambda_1 \cdots \lambda_k] \quad \text{and} \quad \Lambda_\mathrm{II} = [\lambda_{k+1} \cdots \lambda_m].$$

That is to say, $\lambda_i = \alpha p_i$ for some $\alpha > 0$ and every $i = 1, \ldots, m$.

$\square$

I shall proceed now to solve the "transformation problem" between the profit and the exploitation rates for a Leontief economy. Whereas in a simple Marxian economy of the type discussed in chapter 1 the exploitation rate is uniform, in a Leontief economy need not be so, but can vary from one production process to another. In fact, due to Theorem 6 the exploitation rate in a Leontief economy adopts the form introduced by the following theorem.

THEOREM 6: *In a Leontief economy, if the technology chosen by the firms is a RGD, then the profit rate π of any process $\tilde{\mathbf{x}} \in P^+$ can be obtained out of its the exploitation rate by means of the transformation*

$$\theta(\xi) = \left(\xi^{-1} + \frac{\mathbf{p}\mathbf{x}}{\mathbf{p}\hat{\mathbf{x}} - \mathbf{w}\mathbf{x}}\right)^{-1}.$$

Proof: By definition of exploitation rate,

$$\varepsilon(\tilde{\mathbf{x}}) = \frac{\lambda(\hat{\mathbf{x}}) - \lambda(\mathbf{Bx})}{\lambda(\mathbf{Bx})}$$

$$= \frac{\mathbf{p}\hat{\mathbf{x}} - \mathbf{pBx}}{\mathbf{pBx}}$$

$$= \frac{\mathbf{p}\hat{\mathbf{x}} - \mathbf{wx}}{\mathbf{wx}}.$$

Therefore,

$$\theta[\varepsilon(\tilde{\mathbf{x}})] = \left(\frac{\mathbf{wx}}{\mathbf{p}\hat{\mathbf{x}} - \mathbf{wx}} + \frac{\mathbf{px}}{\mathbf{p}\hat{\mathbf{x}} - \mathbf{wx}}\right)^{-1}$$

$$= \pi(\tilde{\mathbf{x}})$$

$$= \pi. \quad \square$$

The firms' choosing a RGD guarantees the reproduction of the productive cycle but does not guarantee full employment, only the possibility of satisfying the demand of those that turn out to be employed. In order to close the present book, I will show that full employment socially and historically presupposes the existence of sufficiently large stocks of capital and wage goods. To this effect I will introduce a new primitive concept, which is the set of all possible distributions of all the available social labor power. This is determined by the technologies possible for the economy as a whole: the technical properties of the possible technologies constrain the possible ways in which all the workers can be allocated among the different production processes. Hence, for any Leontief technology X we may think of the set of all possible allocations of labor power as a subset L of the whole set $\mathcal{L}$ of labor inputs of X (chapter 5, Definition 3). One condition that this set must satisfy is that the vector of labor inputs of any labor process, if properly scaled, must be in L. Clearly, if the global process $\tilde{\mathbf{x}}$ requires more labor than that represented by any of the elements of S, then it cannot be operated by the firms. But the firms can (in fact, they will) operate a global process whose vector of labor inputs represents less labor than any element of S, unless the accumulated vector of initial holdings is sufficiently large. If this is the case, and the global decision of the firms is a RGD under a feasible price system, then the economy can sustain itself indefinitely. The clue to full employment lies, of course, in the scale of the economy. Notice also that there is no international trade in this model, and so all the "contradictions" that might arise are solved within this perfectly closed economy.

NOTES

[1]   From now on, I shall ocassionally refer to this work with the letter *C*. All quotations from this work are taken from the Penguin edition (1976, 1978, 1981). Since there are very different editions, instead of referring to page numbers in making quotations I refer to the book, part, chapter and section, in that order. Thus, for instance, an expression like '*C*1, p1, ch2, s3' denotes the third section of the second chapter, first part of book one.

[2]   *C*1, p1, ch1, s1.

[3]   *Ibid.* The italics are mine.

[4]   *Ibid.* My italics.

[5]   *Ibid.*

[6]   *Ibid.*

[7]   *Ibid.* The italics are mine.

[8]   *C*1, p1, ch1, s2.

[9]   *Ibid.* My italics.

[10]   *Ibid.* My italics.

[11]   *Ibid.*

[12]   *C*1, p1, ch1, s3. My italics.

[13]   *Ibid.*

[14]   *Ibid.* My italics.

[15]   *Ibid.*

[16]   *C*1, p3, ch7, s1.

[17]   See Corollary 2 and Theorem 7 in Kemp and Kimura (1978), pp. 8-9. The Hawking-Simon condition appears as this theorem.

[18]   *C*1, p3, ch9, s1.

[19]   *C*1, p2, ch8, s1.

[20]   Morishima (1973), p. 85.

[21]   Morishima (1973), p. 86.

22 Böhm-Bawerk (1896). I follow the Spanish version published in Argentina in 1974.

NOTES TO CHAPTER 2

1 *C*3, p2, ch10.

2 *C*3, p2, ch9.

3 *Ibid.*

4 *Ibid.*

5 Morishima (1973), pp. 72-74.

6 *C*3, p2, ch10. My italics.

7 *Ibid.*

8 *C*3, p4, ch18.

9 *Ibid.* The italics are mine.

10 *C*3, p6, ch37. My italics.

11 *C*3, p7, ch49.

12 *C*3, p7, ch51.

13 Böhm-Bawerk (1974), p. 57.

14 *Ibid.*, p. 54.

15 *Ibid.*

16 *Ibid.*, p. 58.

17 *Ibid.*, p. 67.

18 *Ibid.*, p. 77.

19 Nuti (1974), p. 43.

20 *Ibid.*

21 See pp. 8 and 9.

22 Cameron (1952), p. 193.

23 Morishima and Seton (1961), p. 204.

24 *Ibid.*

25 *Op. cit.*, p. 209.

26 Cf. pp. 297 and 298.

[27] Morishima (1973). Many of the mathematical techniques used by Morishima in this book were developed mainly in the fifties by mathematical economists such as Samuelson, Arrow and Koopmans. See for instance Koopmans (1951).

[28] See Theorems 1 and 2 of chapter 1.

[29] See p. 131.

[30] See Morishima (1973), p. 173.

[31] Morishima (1974), p. 618.

[32] The convexity assumptions have been used to prove the existence of equilibria in neoclassic economics (See Debreu (1956, 1959)). Roemer uses the convexity assumption in order to derive the existence of what he calls Marxian reproducible equilibria (see Roemer (1981)).

[33] Roemer (1981), p. 38.

[34] In fact, Professor H. Scarf has developed important results on the problem of building a theory of (neoclassical) equilibrium on non-convex finitistic assumptions. See Scarf (1981a, 1981b).

NOTES TO CHAPTER 3

[1] See Enderton (1972), p. 75.

[2] See p. 104.

[3] See Enderton (1972), p. 79.

[4] For a precise definition of decidability see Boolos and Jeffrey (1980).

[5] Which can be seen, for instance, in Suppes (1972).

[6] Wallace and Findlay (1975), p. 50.

[7] See the Susätz to §31. Wallace and Findlay (1975), p. 51.

[8] For a systematic study of the relationships between the philosophy of Hegel and that of Aristotle, see Mure (1970).

[9] See Suarez (1960), Disputation XL, Part VI, §5.

[10] See Brown (1984), pp. 153, n. 12. The other "labyrinth" is the problem of reconciling God's foreknowledge with human freewill.

[11] For a definition of Archimedean, regular, positive, ordered, local semigroup, see Krantz et al. (1971), p. 44.

NOTES TO CHAPTER 4

[1] Wallace and Findlay (1971), §79, p. 113.

2   Wallace and Findlay (1971), §161Z, p. 224.

3   Findlay (1958), pp. 70-1.

4   Findlay (1958), p. 57.

5   Findlay (1958), p. 74.

6   Findlay (1958), p. 75.

7   For a view of idealization as isolation see Mäki (1991).

8   Findlay (1958), p. 77.

9   Findlay (1958), p. 71-2.

10   Findlay (1958), pp. 72-3. The first italics are mine.

11   Miller and Findlay (1971), §382, p. 15.

12   Miller and Findlay (1971), §381, p. 8.

13   *Wissenschaft der Logik*, p. 44; quoted by Findlay (1958), p. 152.

14   Findlay (1958), p. 32.

15   See "Materialism and Matter in Marxism-Leninism", in McMullin (1978).

16   In the same Postface to the Second Edition.

17   Elster says: "I find it hard to believe that Marx would have come to accept the laws of dialectics had he put his mind to them". See pp. 42-3.

18   Elster (1985), p. 37.

19   Dussel (1990), p. 404. The translation is mine.

20   *On the Soul*, 430ª 10-25. See Barnes (1984), volume I. The italics are mine.

21   Aquinatis (1886), p. 455. The translation is mine.

22   In the sections A. Consciousness and B. Self-Consciousness. See Westphal (1989), pp. 154 ff.

23   *Schelling Werke*, v. V, p. 198. My translation. I owe to Dussel (1990) his making me aware of Schelling's criticism of Hegel.

24   *Schelling Werke*, Book III, Lesson XII.

25   Plotinus, *Ennead* V, 4.

26   Cfr. Leclerc (1972), p. 66.

27   *Holy Bible* (The New King James Version). Exodus 3:14-15.

28   Kaufmann (1972), p. 21.

29   See Avineri (1972), chapter 2.

[30]   See Dickey (1987) for a thorough study of the historical, political and theological context in eighteenth century Würtemberg, the land where Hegel was born and where he grew up.

[31]   See Avineri (1972).

[32]   See Waszek (1988).

[33]   See Wood (1990).

[34]   See Stern (1990).

[35]   See Westphal (1989).

[36]   Actually, this is what Stern (1990) does in connection with the *Phenomenology of Spirit*. See pp. 43-54.

[37]   Stern (1990), pp. 40-1.

[38]   This quotation is taken from an unpublished paper dated by Van Heijenoort in 1943 under the pseudonym of Alex Barbon. See Van Heijenoort (1943) in the bibliography.

[39]   Van Heijenoort, op. cit.

[40]   *Ibid.*

[41]   I follow here the German version (1974). Whenever I deem it important, I provide the original German expressions together with their translation.

[42]   Marx and Engels (1974), p. 630.

[43]   The value $(2000/g)^{1/2}$ is obtained by setting $-\frac{1}{2}gt^2 + 1000 = 0$ (which is the position of the particle at the end of the motion), and solving for t.

[44]   Nowak (1980), p. 95.

[45]   Nowak (1980), p. 29.

[46]   Nowak (1980), p. 9.

[47]   For a detailed, albeit a rather schematic presentation of this process in connection with MTV, see Hamminga (1990).

[48]   See Koopmans (1951) and Leontief (1941).

NOTES TO CHAPTER 5

[1]   Morishima (1973, 1974), Okishio (1963), Roemer (1980, 1981).

[2]   Marx (1970), p. 29.

[3]   Marx (1970), p. 34.

[4]   Marx (1970), p. 45.

[5]   *Ibid.*

[6] In C1, p1, ch1, s3.

[7] See Rubin (1972).

[8] Rubin (1972), p. 139.

[9] Rubin (1972), pp. 139-140.

[10] See Krause (1979, 1980, 1981) in the Bibliography.

[11] Cf. equation 4 of chapter 1. The proviso therein does not apply in the present case, i.e. the vectors $\underline{x}_i^T$ are of dimension m.

[12] More precisely, we are requiring $x \geq 0$ and

$$\underline{x} \geq 0 \Rightarrow x \geq 0 \Leftrightarrow \bar{x} \geq 0$$

for every $\langle x, \underline{x}, \bar{x} \rangle \in P_0$.

[13] C1, p3, ch7, s2.

[14] *Ibid.*

[15] *Ibid.*

[16] See Kemp and Kimura (1978), p. 3.

NOTES TO CHAPTER 6

[1] See pp. 73-74.

[2] Roemer (1981), p. 73.

[3] A proof of the nonemptyness of $A_f(p, c)$ is provided in Lemma 2 of the next chapter.

NOTES TO CHAPTER 8

[1] Samuelson (1963), p. 233.

[2] It would be more correct to write $[L(\bar{x})^T, X(\bar{x})^T, \bar{x}^T]$. I ask the forgiveness of the reader for this little notational abuse.

BIBLIOGRAPHY

Álvarez, F., "Estructura y función de la ley del valor: un principio guía para la economía política" in Álvarez, S., F. Broncano, and M. A. Quintanilla (eds.), *Actas: I Simposio Hispano-Mexicano de Filosofía*, vol. 1: *Filosofía e Historia de la Ciencia*, Ediciones Universidad de Salamanca, Salamanca, 1986.

Aquinatis, S. T., *Opuscula philosophica et theologica*, Tiferni Tiberini, Castello, 1886.

Arrow, K. J. and Debreu G., "Existence of an Equilibrium for a Competitive Economy" in *Econometrica* 22, 1954.

Avineri, S., *Hegel's Theory of the Modern State*, Cambridge UP, Cambridge, 1972.

Balzer, W., Moulines, C. U. and Sneed, J. D., *An Architectonic for Science*, D Reidel, Dordrecht, 1987.

Barnes, J. (ed.), *The Complete Works of Aristotle*, Princeton UP, Princeton, (1984).

Böhm-Bawerk, E., "The Conclusion of the Marrxian System", several editions.

Boolos, G., and Jeffrey, R., *Computability and Logic*, Cambridge UP, Cambridge, 1980.

Bourbaki, N. (pseud.), *Elements of Mathematics: Theory of Sets*, Addison–Wesley, Reading, 1968.

Brown, S., *Leibniz*, University of Minnesota Press, Minneapolis, 1984.

Brzeziński, J., F. Coniglione, Theo A. F. Kuipers and L. Nowak, (eds.) *Idealization I: General Problems*, Rodopi, Amsterdam, 1990.

Caldwell, B. (ed.), *Appraisal and Criticism in Economics*, Allen & Unwin, Boston, 1984.

Cameron, B., "The Labour Theory of Value in Leontief Models" in *The Economic Journal* 62, 1952.

Debreu, G., "Market Equilibrium" in *Proceedings of the National Academy of Sciences* 42, 1956.

213

————, *Theory of Value*, Yale UP, New Haven, 1959.

Dickey, L., *Hegel. Religion, Economics and the Politics of Spirit (1770-1807)*, Cambridge UP, New York, 1987.

Dussel, E., *El último Marx y la liberación latinoamericana*, Siglo XXI, Mexico, 1990.

Elster, J., *Making Sense of Marx*, Cambridge UP, Cambridge, 1985.

Enderton, H. B., *A Mathematical Introduction to Logic*, Academic Press, New York, 1972.

Engels, F., *Anti-Dühring*, several editions.

————, *Dialectics of Nature*, several editions.

Findlay, J. N., *Hegel: A Re-Examination*, Collier Books, New York, 1958.

Gale, D., *The Theory of Linear Economic Models*, McGraw-Hill, New York, 1960.

Gantmacher, F. R., *Applications of the Theory of Matrices*, Interscience Publishers, New York, 1959.

García de la Sienra, A., "Elementos para una reconstrucción lógica de la teoría del valor de Marx" in *Crítica* 35, 1980.

————, "The Basic Core of the Marxian Economic Theory" in W. Stegmüller, W. Balzer, W. Spohn, (ed.), *Philosophy of Economics*, Springer-Verlag, Heidelberg, 1982.

————, "Axiomatic Foundations of the Marxian Theory of Value", *Erkenntnis* 29, 1988.

Georgescu-Roegen, N., "Leontief's System in the Light of Recent Results" in *The Review of Economics and Statistics* 32, 1950.

Gracia, J. J. E., *Individuality. An Essay on the Foundations of Metaphysics*, SUNY Press, Albany, 1988.

Hamminga, B., "The Structure of Six Transformations in Marx's *Capital*" in Brzeziński *et al.* (1990).

Hegel, G. W. F., *Wissenschaft der Logik*, Surhkamp, Stuttgart, 1986.

————, *Phenomenology of Spirit*, Clarendon Press, Oxford, 1977.

Holy Bible (The New King James Version), American Bible Society, New York, 1982.

Kakutani, S., "A Generalization of Brouwer's Fixed Point Theorem" in *Duke Mathematical Journal*, 8 (1941).

Kaufmann, W., "The Hegel Myth and Its Method" in McIntyre, A. (ed.), *Hegel*, Anchor-Doubleday, New York, 1972.

Kemp, M. C. and Y. Kimura, *Introduction to Mathematical Economics*, Springer-Verlag, New York, 1978.

Koopmans T. C. (ed.), *Activity Analysis of Production and Allocation*, John Wiley & Sons, New York, 1951.

Kosok, M., "The Formalization of Hegel's Dialectical Logic" in *International Philosophical Quarterly*, vol. VI, no. 4 (1966).

Krantz, D. H., Luce, R. D, Suppes, P., and Tversky, A., *Foundations of Measurement I*, Academic Press, New York, 1971.

Krause, U., *Geld und abstrakte Arbeit*, Campus Verlag, 1979. English edition: *Money and Abstract Labour*, Verso, London, 1982.

_____, "Abstract Labour in General Joint Systems" in *Metroeconomica* 32, 1980.

_____, "Heterogeneous Labour and the Fundamental Marxian Theorem" in *Review of Economic Studies* 48, 1981.

Landau, E., *Foundations of Analysis*, Chelsea, New York, 1966.

Leclerc, I., *The Nature of Physical Existence*, George Allen & Unwin, London, 1972.

Lenin, V. I., *Materialism and Empirio-Criticism*, Foreign Languages Press, Peking, 1972.

Leontief, W. W., *The Structure of American Economy 1919-1929*, Harvard UP, Cambridge, 1941.

Lobkowicz, N., "Materialism and Matter in Marxism-Leninism" in McMullin, E. (ed.), *The Concept of Matter in Modern Philosophy*, University of Notre Dame Press, Notre Dame, 1978.

Mäki, U., "On the Method of Isolation in Economics", forthcoming in Dilworth, C. (ed.), *Intelligibility in Science (Poznan Studies in the Philosophy of the Sciences and the Humanities)*.

Marx, K., *A Contribution to a Critique of Political Economy*, International Publishers, New York, 1970.

_____ and Engels, F., *Werke*, Band 13, Dietz Verlag, Berlin, 1974.

_____, *Capital*, Penguin Books, Harmondsworth; v. 1, 1976; v. 2, 1978; v. 3, 1981.

McLane S. and Birkhoff, G., *Algebra*, Macmillan, New York, 1967.

Miller, A. V. and Findlay, J. N., *Hegel's Philosophy of Mind*, Clarendon, Oxford, (1971).

Morishima, M., *The Theory of Economic Growth*, Clarendon UP, Oxford, 1969.

_____, *Marx's Economics*, Cambridge UP, Cambridge, 1973.

_____, "Marx in the Light of Modern Economic Theory" in *Econometrica* 42, 1974.

_____ and Seton, F., "Aggregation in Leontief Matrices and the Labour Theory of Value" in *Econometrica* 29, 1961.

Mure, G. R. G., *An Introduction to Hegel*, Oxford UP, Oxford, 1970.

Nikaido, H., *Convex Structures and Economic Theory*, Academic Press, New York, 1968.

Nowak, L. : *The Structure of Idealization*, D Reidel, Dordrecht, 1980.

Nuti, D. M., *V. K. Dmitriev: Economic Essays*, Cambridge UP, Cambridge, 1974.

Okishio, N., "A Mathematical Note on Marxian Theorems" in *Weltwirtschaftliches Archiv* 91, 1963.

Robinson, A., *Non-Standard Analysis*, North Holland, Amsterdam, 1961.

Roemer, J. E., "A General Equilibrium Approach to Marxian Economics" in *Econometrica* 48, 1980.

_____, *Analytical Foundations of Marxian Economic Theory*, Cambridge UP, Cambridge, 1981.

Rubin, I. I., *Essays on Marx's Theory of Value*, Black & Red, Detroit, 1972.

Samuelson, P., "Discussion", *American Economic Review Papers and Proceedings*, May (1963). Reprinted in Caldwell (1984).

Scarf, H. E., "Production Sets with Indivisibilities-Part I: Generalities" in *Econometrica* 49, 1981a.

_____, "Production Sets with Indivisibilities-Part II: The Case of Two Activities" in *Econometrica* 49, 1981b.

Schelling Werke, Schroeter, Munich, 1958.

Sneed, J. D., *The Logical Structure of Mathematical Physics*, D. Reidel, Dordrecht, 1971.

Stern, R., *Hegel, Kant and the Structure of the Object*, Routledge, London, 1990.

Suarez, F., *Disputaciones metafísicas*, Gredos, Madrid, 1960.

Suppes, P., *Axiomatic Set Theory*, Dover, New York, 1972.

_____, *Introduction to Logic*, D. van Nostrand, New York, 1957.

_____, *Theoretical Structures in Science* (Preliminary Draft), Manuscript, July 1984.

Van Heijenoort, J., "On Marx's Method in *Capital*", Manuscript, 1943.

Von Bortkiewicz, L., "Zur Berichtigung der grundlegenden theoretischen Konstruktion von Marx im III. Band des *Kapitals*" in *Jarbücher für Nationalökonomie und Statistik* 34, 1907.

Wallace, W. and Findlay, J. N., *Hegel's Logic*, Clarendon Press, Oxford, 1971.

Waszek, N., *The Scottish Enlightenment and Hegel's Account of 'Civil Society'*, Kluwer, Dordrecht, 1988.

Westphal, K. R., *Hegel's Epistemological Realism*, Kluwer, Dordrecht, 1989.

Wood, A. W., *Hegel's Ethical Thought*, Cambridge UP, Cambridge, 1990.

NAME INDEX

219

SUBJECT INDEX

106. K. Kosík, *Dialectics of the Concrete*. A Study on Problems of Man and World. [Boston Studies in the Philosophy of Science, Vol. LII] 1976
ISBN 90-277-0761-8; Pb 90-277-0764-2
107. N. Goodman, *The Structure of Appearance*. 3rd ed. with an Introduction by G. Hellman. [Boston Studies in the Philosophy of Science, Vol. LIII] 1977
ISBN 90-277-0773-1; Pb 90-277-0774-X
108. K. Ajdukiewicz, *The Scientific World-Perspective and Other Essays, 1931-1963.* Translated from Polish. Edited and with an Introduction by J. Giedymin. 1978
ISBN 90-277-0527-5
109. R. L. Causey, *Unity of Science*. 1977　　　　　　　ISBN 90-277-0779-0
110. R. E. Grandy, *Advanced Logic for Applications*. 1977　　ISBN 90-277-0781-2
111. R. P. McArthur, *Tense Logic*. 1976　　　　　　　　ISBN 90-277-0697-2
112. L. Lindahl, *Position and Change*. A Study in Law and Logic. Translated from Swedish by P. Needham. 1977　　　　　　　ISBN 90-277-0787-1
113. R. Tuomela, *Dispositions*. 1978　　　　　　　ISBN 90-277-0810-X
114. H. A. Simon, *Models of Discovery and Other Topics in the Methods of Science.* [Boston Studies in the Philosophy of Science, Vol. LIV] 1977
ISBN 90-277-0812-6; Pb 90-277-0858-4
115. R. D. Rosenkrantz, *Inference, Method and Decision*. Towards a Bayesian Philosophy of Science. 1977　　　ISBN 90-277-0817-7; Pb 90-277-0818-5
116. R. Tuomela, *Human Action and Its Explanation*. A Study on the Philosophical Foundations of Psychology. 1977　　　　　　　ISBN 90-277-0824-X
117. M. Lazerowitz, *The Language of Philosophy*. Freud and Wittgenstein. [Boston Studies in the Philosophy of Science, Vol. LV] 1977
ISBN 90-277-0826-6; Pb 90-277-0862-2
118. Not published
119. J. Pelc (ed.), *Semiotics in Poland, 1894-1969*. Translated from Polish. 1979
ISBN 90-277-0811-8
120. I. Pörn, *Action Theory and Social Science*. Some Formal Models. 1977
ISBN 90-277-0846-0
121. J. Margolis, *Persons and Mind*. The Prospects of Nonreductive Materialism. [Boston Studies in the Philosophy of Science, Vol. LVII] 1977
ISBN 90-277-0854-1; Pb 90-277-0863-0
122. J. Hintikka, I. Niiniluoto, and E. Saarinen (eds.), *Essays on Mathematical and Philosophical Logic*. 1979　　　　　　　ISBN 90-277-0879-7
123. T. A. F. Kuipers, *Studies in Inductive Probability and Rational Expectation*. 1978
ISBN 90-277-0882-7
124. E. Saarinen, R. Hilpinen, I. Niiniluoto and M. P. Hintikka (eds.), *Essays in Honour of Jaakko Hintikka on the Occasion of His 50th Birthday*. 1979
ISBN 90-277-0916-5
125. G. Radnitzky and G. Andersson (eds.), *Progress and Rationality in Science*. [Boston Studies in the Philosophy of Science, Vol. LVIII] 1978
ISBN 90-277-0921-1; Pb 90-277-0922-X
126. P. Mittelstaedt, *Quantum Logic*. 1978　　　　　ISBN 90-277-0925-4
127. K. A. Bowen, *Model Theory for Modal Logic*. Kripke Models for Modal Predicate Calculi. 1979　　　　　　　ISBN 90-277-0929-7
128. H. A. Bursen, *Dismantling the Memory Machine*. A Philosophical Investigation of Machine Theories of Memory. 1978　　　　ISBN 90-277-0933-5

129. M. W. Wartofsky, *Models*. Representation and the Scientific Understanding. [Boston Studies in the Philosophy of Science, Vol. XLVIII] 1979
ISBN 90-277-0736-7; Pb 90-277-0947-5
130. D. Ihde, *Technics and Praxis*. A Philosophy of Technology. [Boston Studies in the Philosophy of Science, Vol. XXIV] 1979   ISBN 90-277-0953-X; Pb 90-277-0954-8
131. J. J. Wiatr (ed.), *Polish Essays in the Methodology of the Social Sciences*. [Boston Studies in the Philosophy of Science, Vol. XXIX] 1979
ISBN 90-277-0723-5; Pb 90-277-0956-4
132. W. C. Salmon (ed.), *Hans Reichenbach: Logical Empiricist*. 1979
ISBN 90-277-0958-0
133. P. Bieri, R.-P. Horstmann and L. Krüger (eds.), *Transcendental Arguments in Science*. Essays in Epistemology. 1979   ISBN 90-277-0963-7; Pb 90-277-0964-5
134. M. Marković and G. Petrović (eds.), *Praxis*. Yugoslav Essays in the Philosophy and Methodology of the Social Sciences. [Boston Studies in the Philosophy of Science, Vol. XXXVI] 1979   ISBN 90-277-0727-8; Pb 90-277-0968-8
135. R. Wójcicki, *Topics in the Formal Methodology of Empirical Sciences*. Translated from Polish. 1979   ISBN 90-277-1004-X
136. G. Radnitzky and G. Andersson (eds.), *The Structure and Development of Science*. [Boston Studies in the Philosophy of Science, Vol. LIX] 1979
ISBN 90-277-0994-7; Pb 90-277-0995-5
137. J. C. Webb, *Mechanism, Mentalism and Metamathematics*. An Essay on Finitism. 1980   ISBN 90-277-1046-5
138. D. F. Gustafson and B. L. Tapscott (eds.), *Body, Mind and Method*. Essays in Honor of Virgil C. Aldrich. 1979   ISBN 90-277-1013-9
139. L. Nowak, *The Structure of Idealization*. Towards a Systematic Interpretation of the Marxian Idea of Science. 1980   ISBN 90-277-1014-7
140. C. Perelman, *The New Rhetoric and the Humanities*. Essays on Rhetoric and Its Applications. Translated from French and German. With an Introduction by H. Zyskind. 1979   ISBN 90-277-1018-X; Pb 90-277-1019-8
141. W. Rabinowicz, *Universalizability*. A Study in Morals and Metaphysics. 1979
ISBN 90-277-1020-2
142. C. Perelman, *Justice, Law and Argument*. Essays on Moral and Legal Reasoning. Translated from French and German. With an Introduction by H.J. Berman. 1980
ISBN 90-277-1089-9; Pb 90-277-1090-2
143. S. Kanger and S. Öhman (eds.), *Philosophy and Grammar*. Papers on the Occasion of the Quincentennial of Uppsala University. 1981   ISBN 90-277-1091-0
144. T. Pawlowski, *Concept Formation in the Humanities and the Social Sciences*. 1980
ISBN 90-277-1096-1
145. J. Hintikka, D. Gruender and E. Agazzi (eds.), *Theory Change, Ancient Axiomatics and Galileo's Methodology*. Proceedings of the 1978 Pisa Conference on the History and Philosophy of Science, Volume I. 1981   ISBN 90-277-1126-7
146. J. Hintikka, D. Gruender and E. Agazzi (eds.), *Probabilistic Thinking, Thermodynamics, and the Interaction of the History and Philosophy of Science*. Proceedings of the 1978 Pisa Conference on the History and Philosophy of Science, Volume II. 1981   ISBN 90-277-1127-5
147. U. Mönnich (ed.), *Aspects of Philosophical Logic*. Some Logical Forays into Central Notions of Linguistics and Philosophy. 1981   ISBN 90-277-1201-8
148. D. M. Gabbay, *Semantical Investigations in Heyting's Intuitionistic Logic*. 1981
ISBN 90-277-1202-6

199. R. Wójcicki, *Theory of Logical Calculi.* Basic Theory of Consequence Operations. 1988 ISBN 90-277-2785-6
200. J. Hintikka and M.B. Hintikka, *The Logic of Epistemology and the Epistemology of Logic.* Selected Essays. 1989 ISBN 0-7923-0040-8; Pb 0-7923-0041-6
201. E. Agazzi (ed.), *Probability in the Sciences.* 1988 ISBN 90-277-2808-9
202. M. Meyer (ed.), *From Metaphysics to Rhetoric.* 1989 ISBN 90-277-2814-3
203. R.L. Tieszen, *Mathematical Intuition.* Phenomenology and Mathematical Knowledge. 1989 ISBN 0-7923-0131-5
204. A. Melnick, *Space, Time, and Thought in Kant.* 1989 ISBN 0-7923-0135-8
205. D.W. Smith, *The Circle of Acquaintance.* Perception, Consciousness, and Empathy. 1989 ISBN 0-7923-0252-4
206. M.H. Salmon (ed.), *The Philosophy of Logical Mechanism.* Essays in Honor of Arthur W. Burks. With his Responses, and with a Bibliography of Burk's Work. 1990 ISBN 0-7923-0325-3
207. M. Kusch, *Language as Calculus vs. Language as Universal Medium.* A Study in Husserl, Heidegger, and Gadamer. 1989 ISBN 0-7923-0333-4
208. T.C. Meyering, *Historical Roots of Cognitive Science.* The Rise of a Cognitive Theory of Perception from Antiquity to the Nineteenth Century. 1989 ISBN 0-7923-0349-0
209. P. Kosso, *Observability and Observation in Physical Science.* 1989 ISBN 0-7923-0389-X
210. J. Kmita, *Essays on the Theory of Scientific Cognition.* 1990 ISBN 0-7923-0441-1
211. W. Sieg (ed.), *Acting and Reflecting.* The Interdisciplinary Turn in Philosophy. 1990 ISBN 0-7923-0512-4
212. J. Karpiński, *Causality in Sociological Research.* 1990 ISBN 0-7923-0546-9
213. H.A. Lewis (ed.), *Peter Geach: Philosophical Encounters.* 1991 ISBN 0-7923-0823-9
214. M. Ter Hark, *Beyond the Inner and the Outer.* Wittgenstein's Philosophy of Psychology. 1990 ISBN 0-7923-0850-6
215. M. Gosselin, *Nominalism and Contemporary Nominalism.* Ontological and Epistemological Implications of the Work of W.V.O. Quine and of N. Goodman. 1990 ISBN 0-7923-0904-9
216. J.H. Fetzer, D. Shatz and G. Schlesinger (eds.), *Definitions and Definability.* Philosophical Perspectives. 1991 ISBN 0-7923-1046-2
217. E. Agazzi and A. Cordero (eds.), *Philosophy and the Origin and Evolution of the Universe.* 1991 ISBN 0-7923-1322-4
218. M. Kusch, *Foucault's Strata and Fields.* An Investigation into Archaeological and Genealogical Science Studies. 1991 ISBN 0-7923-1462-X
219. C.J. Posy, *Kant's Philosophy of Mathematics.* Modern Essays. 1992 ISBN 0-7923-1495-6
220. G. Van de Vijver, *New Perspectives on Cybernetics.* Self-Organization, Autonomy and Connectionism. 1992 ISBN 0-7923-1519-7
221. J.C. Nyíri, *Tradition and Individuality.* Essays. 1992 ISBN 0-7923-1566-9
222. R. Howell, *Kant's Transcendental Deduction.* An Analysis of Main Themes in His Critical Philosophy. 1992 ISBN 0-7923-1571-5

223. A. García de la Sienra, *The Logical Foundations of the Marxian Theory of Value.*
1992   ISBN 0-7923-1778-5

Previous volumes are still available.

KLUWER ACADEMIC PUBLISHERS – DORDRECHT / BOSTON / LONDON

.